# $H_2O$

# HTO

## Toronto's Water from Lake Iroquois to Lost Rivers to Low-flow Toilets

edited by **Wayne Reeves**
and **Christina Palassio**

Coach House Books
Toronto

First edition

Published with the assistance of the Canada Council for the Arts and the Ontario Arts Council. We also acknowledge the financial support of the Government of Ontario through the Ontario Book Publishing Tax Credit Program and the Government of Canada through the Book Publishing Industry Development Program.

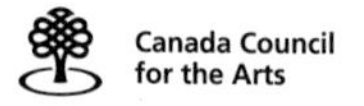


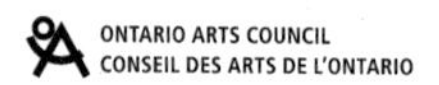

We have deliberately included a diversity of perspectives for this collection in order to reflect a multifaceted Toronto. The opinions expressed in these essays do not necessarily reflect the opinions of the editors or Coach House Books.

Library and Archives Canada Cataloguing in Publication

HTO : Toronto's water from Lake Iroquois to lost rivers to low-flow toilets / edited by Wayne Reeves and Christina Palassio.

ISBN 978-1-55245-208-0

1. Water--Ontario--Toronto. 2. Toronto (Ont.).
I. Reeves, Wayne, 1959- II. Palassio, Christina

TD227.T6H86 2008 553.709713'541 C2008-905637-X

# Contents

**Explorations**



**Directions**

# Introduction

## Bridging the past, present and future of Toronto's water

It's September 2008, and Toronto is towelling off from its wettest summer since record-keeping began at Pearson Airport seventy-one years ago. This rain-soaked summer came hard on the heels of the winter of 2007–08, when we found ourselves within a modest storm of an all-time record for snowfall. Before that, the summer of 2007 had been the driest since 1959, part of a ten-month drought that parched the Greater Toronto Area.

Conspicuous by its presence or absence, water makes us talk. It's not just idle chatter about the weather. It's bound up with concerns about bottled-water bans and broken water mains, flooded basements and flood-plain restoration, downspout disconnections and drugs in drinking water, beach postings and Bisphenol A, water-use reductions and rising water rates.

But while Canadians may take water for granted – only Americans consume more per person than we do, and we pay the least for it – we're also at the forefront of thinking and writing about water issues. Whether the scale is national, continental or global, Canadians are renowned for tackling the big issues of water use, governance and management.[1]

So, why focus only on Toronto?

The local matters. Yes, we need to keep an eye on the larger scene. This was the year, after all, that our federal government helped block the United Nations from recognizing access to water as a basic human right, that Ontario signed on to the Great Lakes–St. Lawrence River Basin Sustainable Water Resources Agreement to control large-scale water diversions, that movement got underway to reopen the Canada-U.S. Great Lakes Water Quality Agreement, that the International Joint Commission made moves to decide how water levels and flows in Lake Ontario and the St. Lawrence River will be regulated. Most of these matters have implications for Toronto, as do national and provincial policies, programs, rules and regulations. But they're not up for discussion here.[2]

The local matters because it's the place people truly care about, and where the action happens, for better or worse. 'Walkerton' has become shorthand for systemic failure

1 See: Maude Barlow and Tony Clarke, *Blue Gold: The Battle Against Corporate Theft of the World's Water* (Toronto: Stoddart, 2002); Marq de Villers, *Water: The Fate of Our Most Precious Resource* (Toronto: McClelland and Stewart, 2003); Karen Bakker, ed., *Eau Canada: The Future of Canada's Water* (Vancouver: UBC Press, 2007); Maude Barlow, *Blue Covenant: The Global Water Crisis and the Coming Battle for the Right to Water* (Toronto: McClelland and Stewart, 2007); Chris Wood, *Dry Spring: The Coming Water Crisis of North America* (Vancouver: Raincoast Books, 2008).

2 We can't ignore one example of how local conditions press up against the big picture, however. The day the Ontario Legislature approved the *Clean Water Act* in 2007, advisories were posted around the Ministry of the Attorney General building at 720 Bay Street warning people not to drink the tap water due to lead contamination from old pipes.

FOUNDATIONS
*Detail from Eldon Garnet's* Time: And A Clock, *on the Queen Street East bridge over the Don. Built in 1911, this is the oldest truss roadway bridge spanning a waterway in Toronto.*

in providing safe drinking water. With the opening of the Walkerton Clean Water Centre in 2005, it is now also a place where we can learn how to avoid repeating past mistakes. Local governments are spending $15 billion annually to protect and restore the Great Lakes and the St. Lawrence – suggesting that only at the local level can citizens affect real change.

FACING PAGE

FOUNDATIONS
*Featured in dozens of print and TV ads and photoblogs, the Humber Pedestrian and Cycling Bridge (1995) is perhaps our most iconic lakeshore structure. A closer look reveals a wealth of Aboriginal symbols. A flock of Thunderbirds, bracing the bridge's arches, rises into the Sky World. Turtle, on the bridge parapets, carries the Middle World on its back and ensures that people cross in peace and safety. Serpent, spirit of great strength, guards the waters of the Middle World and the bridge abutments leading to the Under World given to it by the Creator. Guardian Faces watch both ways over the ancient Toronto Carrying Place, as the modern-day Waterfront Trail intersects with it and our most historic waterway.*

TRANSFORMATIONS
*In John George Howard's 1834 painting of the third parliament buildings of Upper Canada at Front and Simcoe streets, we think the focal point is actually the culvert-cum-bridge over Russell Creek. The stream was soon completely erased from the landscape, its burial fuelled by the arrival of brick sewers upstream in 1835, and by railway-related lake-filling to the south twenty years later.*

Toronto is an instructive case study in how localities relate to water. Drained by a half-dozen major watersheds, cut by a network of deep ravines, fronting on a Great Lake and home to huge drinking-water, wastewater and flood-control infrastructures, Toronto is a city dominated by water. The trend of fettering Toronto's water and putting it underground, and of degrading that which still flows on the surface, has recently been countered by persistent citizen-led efforts to recall, rethink and restore our communal aqua.

Watersheds surface frequently in *HTO* as a geographic unit that's a useful, insistent rebuttal to local introspection. This unit is the basis for the organization of Ontario's conservation authorities, which began to manage water in the Toronto region in 1946. More importantly, though, water is a flow resource, constantly moving across political boundaries and connecting communities. Toronto is part of a nested set of drainage areas: our watersheds (2,135 km$^2$) lie within the Lake Ontario Basin (83,000 km$^2$), which in turn is part of the Great Lakes Basin (766,000 km$^2$). Water may be used, abused and improved locally, but the impacts are felt far away, in other people's backyards. The future of Toronto's water is not ours alone. Despite what we do here, it matters what happens in Thunder Bay and Thornhill. And what we do affects Kingston and Quebec City. We're all downstream.

So, while remaining mindful of our watershed and basin contexts, talk of water in *HTO* remains primarily grounded in the City of Toronto. It's ultimately where we'll have to take responsibility for our past and present actions, and where we should get credit for our achievements. Sometimes we're both castigated and praised by the same organization. In 2006, Sierra Legal (now Ecojustice) gave Toronto a C in *The Great Lakes Sewage Report Card* because of our raw-sewage discharges, the frequency of combined sewer overflows (CSOs) and the average yearly volume of those overflows. In 2008, Ecojustice held

Toronto up as a Great Lakes model for using green infrastructure to mitigate CSOs and stormwater.[3] We're a tiny fraction of the Great Lakes Basin, yet our capacity for doing good or ill – in terms of physical impact or policy influence – is so much larger. Toronto is nowhere near perfect on the water front, but in many ways we're heading there.[4]

The words that form part of Eldon Garnet's 1996 public art project, *Time: And A Clock*, on the Queen Street bridge over the Don, neatly express a view that runs through this book. 'This river I step in is not the river I stand in' speaks to fluidity and change. And for us, it's more than a metaphor that cleverly takes advantage of its stream setting. Time for water is both linear and cyclical. We want to know where that river came from, why and how it was changed, what it looks like today and what needs to be done to make it better tomorrow. These questions form the backbone of *HTO*'s four sections: Foundations, Transformations, Explorations and Directions.

The Foundations section lays out the physical basis for Toronto's water and traces our earliest engagement with it. Using the Town of York's founding in 1793 as a starting point, the Transformations section examines some of the processes that gave our water an engineered character. The Explorations section looks at how Torontonians are connecting with water in some of its many guises. The Directions section outlines potential courses toward a better future for water in Toronto.

This four-fold structure (which quickly breaks down, as most of the twenty-nine essays touch on all four themes) can be framed against a broad sense of how Torontonians have related to water over time. Water becomes a lens to look at the changing interplay between nature and culture in Toronto, calling into question our past, present and future engagement with water in its diverse forms. Three cycles – the natural, the engineered and the conserver – sum up our shifting relationship to water.

The natural water cycle existed before European settlers arrived. Precipitation soaked into the ground in a largely forested landscape, nourishing both plants and waterways. Trees and soil intercepted and absorbed much of the rainfall and snowmelt, and a relatively small amount ran off over the land into rivers and lakes. The water was filtered as it slowly

3 The 2008 report is called *Green Cities, Great Lakes*. According to the 2006 report, 906.4 million litres of raw sewage (0.22 percent of total flow) bypassed Toronto's wastewater plants in 2004, while forty to fifty CSOs annually amounted to about nine billon litres of spilled raw sewage (2.14 percent of total flow). Kudos to Toronto for disclosing this data; both York and Durham regions refused to participate in the study.

4 Being at the mouth of six watersheds leaves Toronto especially exposed to 'upstream' pollution. Only 38 percent of developed land in Greater Toronto has some form of stormwater quantity/quality controls. In the Humber watershed – the largest unit directly affecting the City – this figure dips to an alarming 25 percent. See Toronto City Summit Alliance, *Greening Greater Toronto* (Toronto, 2008); Toronto and Region Conservation Authority, *Listen to Your River: A Report Card on the Health of the Humber River Watershed* (Toronto, 2007).

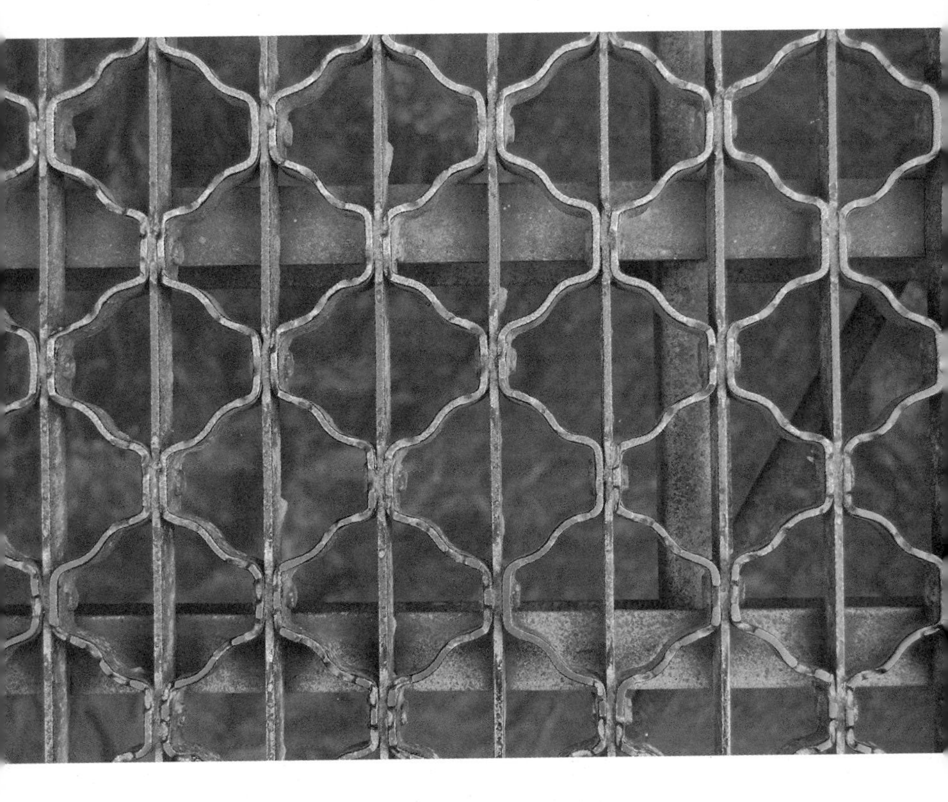

**TRANSFORMATIONS** *After Hurricane Hazel washed out river crossings across the Toronto region in 1954, the 2nd Field Engineer Regiment of the Canadian Military Engineers threw a 'temporary' one-lane Bailey bridge (a temporary bridge usually used in military operations) over the Rouge at Old Finch Avenue. It's still in use, the deck's open structure giving a good view of the river below. The Rouge bridge is the only one of its type left in Toronto's road network.*

passed through the earth to the groundwater table; some remained stored in aquifers, while the rest provided base flow for streams and support for aquatic life. Moisture evaporated from surface waters and transpired from plants into the atmosphere, concentrated into clouds, and, once it fell, the cycle began again. Deliberately or inadvertently, we've spent most of the last 200 years trying to break this cycle. And we've largely succeeded.

The engineered water cycle is the hard result. It developed in Toronto in step with the urbanization process, as we cleared and paved over our watersheds and wetlands and installed infrastructure systems. Precipitation falls mainly onto impervious surfaces, like roofs and roads, picking up pollutants before entering the sewer system. Little rainfall or snowmelt infiltrates the landscape, meaning lots of runoff and not much natural stream recharge. Most stormwater is piped, hot, fast and dirty, into rivers or the lake. Some is combined with sewage and heads to wastewater-treatment plants. The rest of our sewage is carried there by sanitary sewers. The processed wastewater typically goes into the lake as treated effluent, though treatment capacity is overwhelmed during heavy storms, resulting in the CSOs and raw discharges noted above. At the lake, the cycle splits again into two streams: some water evaporates and condenses into clouds; some is purified at water-treatment plants and piped uphill for our consumption.

The engineered water cycle includes more than tap water and poo and the trinity of consumption, elimination and treatment. Our rivers were first impounded by dams and diverted through millraces (a water current that drives a mill wheel) to support Toronto's earliest industries. Then came the issue of efficient drainage, and the need to send our streams underground, hold them behind flood-control dams, or armour their banks and channelize them, all to handle increased stormwater flows. In other cases, water just got in the way of urban-industrial development and was reorganized or eliminated.[5] That's why Toronto has lost 92 percent of its lakeshore marsh acreage since the late eighteenth century. As Metro Toronto chairman Frederick G. Gardiner said of the Don Valley Parkway in the 1950s, 'The problem was that there were two big hills and a narrow-gutted valley. There were railways in it and a river . . . We'll move the railway over a piece. We'll tear down the hill.

5 But don't forget all the playful water features that engineering *added* to the city: swimming pools, splash pads, decorative fountains, ice rinks and the like. The aesthetic and recreational pleasures of Toronto's water will have to wait for another book. Or you can take a look at *Spacing*'s summer 2007 issue on water.

EXPLORATIONS
*As part of the dismantling of the east end of the Gardiner Expressway in 2000–02, generous pedestrian and cycling facilities were added to the north side of Lake Shore Boulevard, including a new bridge connection to the Lower Don Recreational Trail. This modest structure increases access to the Don Valley and the Keating Channel area, as walkers discovered during a Lost Rivers/RiverSides workshop on Canadian Rivers Day, 2008.*

We'll shift the river over a piece, then we can have the highway through here.'[6]

To so casually dispense with nature isn't quite what you see in today's environmental assessments. Gardiner's mindset embodies the core assumptions of the engineered water cycle: that nature is there to be manipulated for human ends, that we can always supply more of what we want and that we have the technology, expertise, cash and willpower to do it. The outcome? Massive, centralized and inflexible infrastructure that's vulnerable to unpredicted changes and hard to adapt to new conditions, like our changed climate. Expensive, too: in 2005, Toronto Water valued its assets at $26.6 billion. At the household level, that mindset is represented by your old bathroom toilet, which sucks up twenty litres of water to move your pee out of sight without you having any idea where it's all gone.

FACING PAGE

DIRECTIONS
*Spring 2008 brought the installation of a temporary artwork on, and downstream of, a now-abandoned 1933 railway bridge over the lower Don. BGL's* Project for the Don River *featured an oversized lifebuoy and a scaled-down version of the* Queen Mary II *that raised questions about rescuing and using the Don, both now and in the future.*

DIRECTIONS
*In 1997, the first single-rib inclined-arch bridge in North America was placed over Mimico Creek. A minor showcase for Spanish engineer/architect Santiago Calatrava (better known in Toronto for the Galleria at BCE Place), the bridge supported the development of the Waterfront Trail across Etobicoke and enabled users to experience the creek in a new way.*

The conserver water cycle challenges the flush-and-forget mentality. It means we're aware of the limitations of the engineered water cycle and we're thinking about how we can get back some of the character and hydrological functions of the natural water cycle. It means using less water whenever we can, extracting more work from the water we do use and stewarding every drop that falls. It's about limiting what we take out of nature, and returning the right amount of higher-quality water back to nature.

If that sounds a bit mystical, it really starts with 'more practical, cautious and reasonable water use,' to quote from Chris Wood's *Dry Spring: The Coming Water Crisis of North America.* Not *BlueTOpia*, but hydro-logical. Some commentators claim that water use could be cost-effectively cut by at least thirty percent using off-the-shelf technologies. It's okay if you jump into the conserver cycle to save some cash on your metered water bill or to get a rebate for that new six-litre toilet. Toronto Water's doing the same thing, spending $75 million on water-efficiency measures to avoid spending $220 million to expand its infrastructure.

Responsibility for making this cycle work will have to be shared broadly. It's not just up to the government and engineers, though public leadership and innovative technology are needed. It's going to take individual action at the household level, community effort in our neighbourhoods and investment by private corporations to reduce water demand,[7] protect our water sources and create more flexible, resilient infrastructure.

6 Quoted in the *Toronto Star*, November 15, 1961.

7 Lower consumption creates other challenges for Toronto Water. After Labatt closed its Toronto brewery in 2005, water rates went up 1 percent to offset the lost revenue.

DON
RIVER

The new landscape of water conservation is fairly subtle: a disconnected downspout leading to a backyard rain barrel, a stormwater pond in a valley-bottom park, permeable paving in the Smart!Centres parking lot, landscaped swales by the roadside, green roofs atop schools, the disappearance of cooling towers on downtown office towers.[8] We'll start by watering our lawns less, and then we'll transform them into gardens that need little or no tap water.

And that, coupled with watching Fido lap water out of the toilet bowl, will lead to more fundamental questions. Do we really need to use our finest drinking water to flush away our wastes, wash our ultra-low-emission cars or soak our local playing fields? We'll shift from doing the same old thing with less water to using the right amount and the right *quality* of water, depending on how and where it's being used.

Which brings us back to the contents of the book. We think that properly informed action should flow from an understanding of how Toronto's water came to be, how and why it changed, and how we experience it. So, alongside instructions on how to grow your own bog garden, you'll also read about the formation of Lake Admiralty, the significance of the Don Valley Brick Works, the power of community involvement and the growing complexity of local water governance, how lost creeks are inspiring new modes of civic design, ecological recovery and watershed awareness, and how water, energy and conservation intersect in the new concept of 'watergy.'

There's lots of scrutiny of the roles being played by the City government in water management. We look at the billion-dollar Wet Weather Flow Management Master Plan and its implementation, given that stormwater is now the chief source of pollution in Toronto's waterways, and how the latest round of climate change is pummelling our rivers and bridges. The disconnect[9] between City policy and procedure is noted in a profile of citizen-created ditch and bog gardens. We end with a lively account of Toronto's water-efficiency and water-conservation pursuits[10] – said to be a tale of 'sparring aquacultures,' but one where the participants still share many elements of the conserver mindset. And that's a water cycle we all need to be pedalling.

**8** This last point is a nod to Enwave's deep-lake water-cooling system, a Canadian invention that had its first large-scale application in Toronto (2004). Demonstrating the role engineers and big technology can play in the conserver water cycle, DLWC reduces the electricity demands of building cooling systems by 90 percent when compared to conventional chillers.

**9** A gentle reminder that downspout disconnections are now mandatory for 120,000 households in the combined sewer area of Toronto. Eventually, another 230,000 homes connected to storm sewers will also need to be disconnected.

**10** According to Statistics Canada, the Toronto Census Metropolitan Area in 2006 did pretty well compared to the rest of Canada in terms of water-conservation devices and practices. We were above the national average for water-saving showerheads (56 percent versus 53 percent), toilets (38 percent versus 34 percent) and sprinkler timers (26 percent versus 25 percent), but below – in fact, at the bottom – for rain barrel or cistern use (7 percent versus 11 percent).

# Foundations

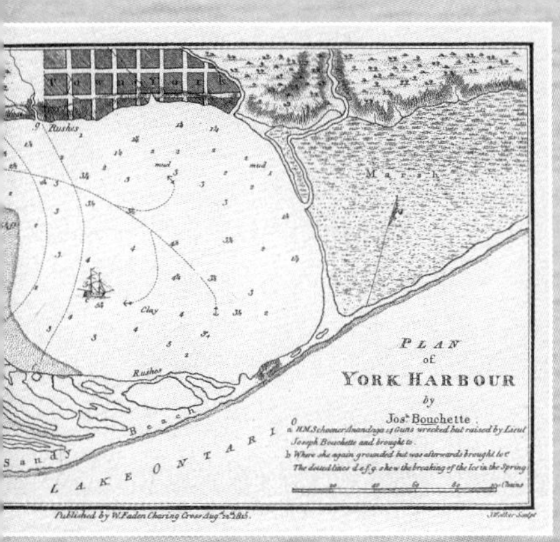
PLAN
of
YORK HARBOUR
by
Jos[h]. Bouchette.
a H.M. Schooner Onondaga 14 Guns wrecked but raised by Lieut Joseph Bouchette and brought to.
b Where she again grounded but was afterwards brought to.
The dotted lines d.e.f.g. shew the breaking of the Ice in the Spring.
Chains
Rushes
Clay
Marsh
Beach
Sandy
LAKE ONTARIO
Published by W. Faden Charing Cross Aug[t]. 12[th]. 1815.

Ed Freeman

# Formed and shaped by water: Toronto's early landscape

Toronto's site was created by water. The action of water formed the city's protected harbour, and water routes connected hinterland resources to the town. From the city's earliest beginnings as a settlement, water was sought from springs and wells for domestic use and to quench the thirst of the citizens. In 1850, when the population of Toronto was about 30,000, the Ontario Brewery and Turner's Brewery on Front Street each had a weekly capacity of 1,500 gallons – or 24,000 pints of beer – and were only two of many breweries within the city. Water provided food in the form of fish and marshland birds, and powered the first mills on the Humber and Don rivers. It was also the medium in which materials were deposited to form the rock beneath Toronto. In 1882, the Copeland Brewery drilled a well at the foot of Parliament Street in search of clean water or natural gas, and in 1890, another well at the intersection of the Queensway and Windermere Avenue in Swansea sought natural gas for the Ontario Bolt Works factory. Both wells failed to produce what was desired, but they revealed layer after layer of water-deposited materials that had turned to shale and limestone before finding their end in the weathered surface of the Precambrian Canadian Shield.

Toronto's oldest known association with water lies 379 metres below ground – roughly the height of a 125-storey building – where the aforementioned well encountered Precambrian limestone at the Bolt Works site. This limestone was formed more than 1.4 billion years ago in an ancient sea. Much, much later, as a result of plate tectonic movement and the metamorphism of this old limestone by heat and pressure, the land slowly rose as a mountain chain. Then, between 470 and 448 million years ago (in the Ordovician period), these mountains eroded into an adjacent subtropical sea, and younger layers of clays and lime-rich muds buried the Precambrian rocks. Occasional volcanic eruptions left thin layers of windblown ash on the ocean bottom. Still later, seasonal hurricanes churned a shallow subtropical sea, leaving storm channels filled with sands and the skeletal fragments of clams, snails, crinoids and moss animals. This water-laid Ordovician sea bottom, now shale

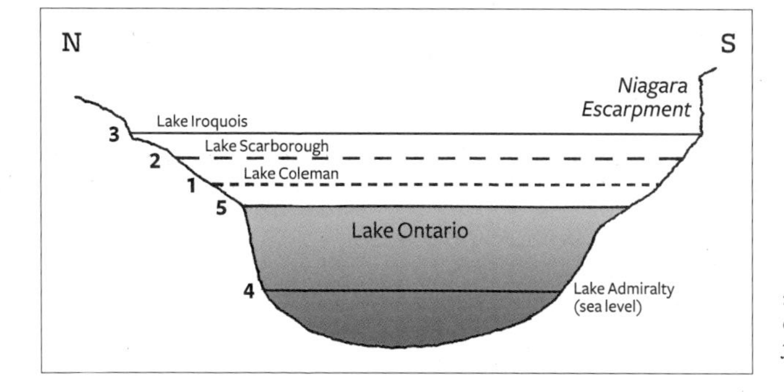


**1** *Some past lake levels in the Lake Ontario Basin over the past 135,000 years.*

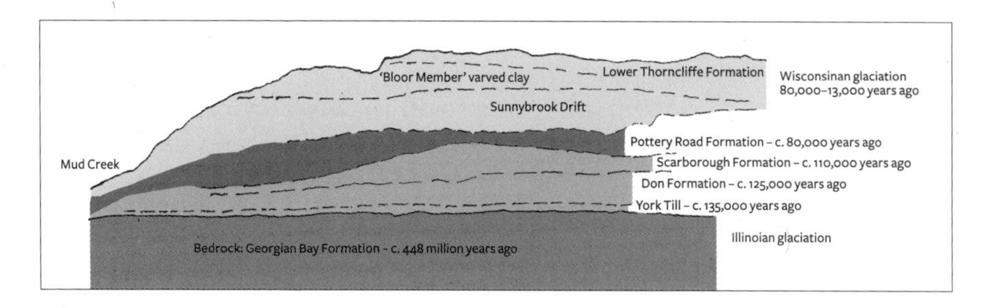


**2** *The deposits of the North Slope at the Don Valley Brick Works.*

and limestone, lies on top of the Precambrian Canadian Shield rocks and forms the bedrock that underlies Toronto, a stable base for many downtown office and condominium towers.

The sequence of bedrock layers nearest the surface of the Ordovician rocks is the Georgian Bay Formation. Building excavations in the downtown area often reveal exposures of these shales, and natural exposures of these old water-laid rocks and their fossil contents have also been uncovered by streams and rivers that have eroded their valleys and ravines. These rocks are visible along the stream banks of the Don, Humber, Mimico, Etobicoke and Credit valleys.

Following the formation of these Ordovician rocks, the upper Great Lakes drained across what is today Toronto, along river channels created 2 to 3 million years ago. This large waterway, called by some the Laurentian River, eroded the surface of the Ordovician layers to create valleys that now lie beneath today's Humber and Don valleys. These old channel bottoms lie 115 metres beneath Lake Ontario offshore from the Humber River, and twenty-five metres below the lake at the mouth of the Don.

Later, glacial ice crept across the area, carrying sand, gravel and boulders eroded from the Canadian Shield materials

A. P. COLEMAN (1852–1939) Arthur Philemon Coleman was born in Lachute, in what is now Quebec, and grew up in the country, where he developed a skill for sketching and painting. His primary interest was in science, however, and he took chemistry, mineralogy, botany and zoology at the University of Breslau, where he obtained a PhD in 1881. He returned to Canada and became professor of geology and natural history at Victoria University in Cobourg, where he worked from 1882 to 1891. In 1891, Victoria University amalgamated with the University of Toronto, and he moved to Toronto to teach at the School of Practical Science until 1901, when he was appointed head of the Department of Geology, a position he held until his retirement in 1922. Coleman travelled widely to observe mountains and glaciers, but he is perhaps best known for his studies of glaciations during the Pleistocene Epoch around Lake Ontario. From 1898 until his death in 1939, he continued to observe, report, sketch, paint and publicize geological events – including those revealed by the deposits within the Don Valley Brick Works. Coleman received many honours and awards, and was elected president of several scientific groups. In addition to hundreds of scientific articles, he wrote two books that deal with the effects of ice: *Ice Ages, Recent and Ancient* and *The Last Million Years*. Coleman is buried in Mount Pleasant Cemetery, where a Precambrian boulder simply inscribed GEOLOGIST marks his grave.

that scoured and buried the Laurentian River channels. After glaciation, there was more subsidence within the glacially filled Laurentian River valleys than on their sides, so today the Don and Humber rivers flow mostly along the more compacted Laurentian channels.

As you're now aware, the sequence and history of glaciations and water levels in the Lake Ontario Basin are complex. A succession of glacial advances and retreats over the last two million years, along with rising and falling lake levels [1], produced a series of deposits that reflect the climate and life at the time of their formation. Most of this record was destroyed by recurring glaciations at Toronto. However, the site of the Don Valley Brick Works reveals the story of the environment and climate of the last 135,000 years [2]. Much of the interpretation of this time period was researched and described by geologist A. P. Coleman as a result of his many visits to the site during excavation of the property.

Beginning in 1889, the Don Valley Brick Works produced a range of clay products from the shale bedrock and the sand and clay sequence lying above it. In 1893, the company sent two boxcars of brick to Chicago for use in the Canada Pavilion at the World's Columbian Exposition, the first World's Fair. The Brick Works won two gold medals for its brick and terra cotta, and the ensuing demand for its products kept it busy for 100 years. Shale was quarried to supply material for red brick; when ground, the shale layers made high-quality plain and specially moulded red bricks. In 1894, the company advertised that it could provide bricks in ten shades of red. The fossils recovered from the shale quarry told the story of its origins as a subtropical sea while, above the shale bedrock, younger deposits provided material for both red and yellow brick.

Chemistry defines the colour of brick. If there is a predominance of iron in the raw material, it bakes red, but if lime is present in abundance, the brick bakes yellow. The near-surface 'blue clay' deposits found over most of southwestern Ontario, washed free from the boulders and gravel of the typical till, contain 10 to 23 percent lime. This blue clay produces the yellow or white bricks that were so abhorrent to Oscar Wilde on his visit to Toronto. (Wilde felt brick should be the good English Georgian red he grew up with.) Many early brickyards found that the top metre of these clay deposits baked red, whereas material below this depth would bake yellow. This 'red-top clay' was the result of rain and snowmelt dissolving the lime from the upper layer as it drained to the water table.

Since 1892, when Coleman began his studies at the Brick Works, a fascinating story of environmental and climate change has emerged. Glacial and river delta deposits above the bedrock there have revealed a marked environmental change in the region, from a glacial period to a climate that was a little warmer than the current one, then on to another glacial period and, lastly, to our present climate. The Brick Works is the only site in the Great Lakes Basin where this entire sequence is revealed [3].

The interglacial story comes from the water-laid deposits of the Don and Scarborough deltas, which were sandwiched between glacier-related deposits. Immediately on top of the bedrock at the Don Valley Brick Works, there is a metre-thick layer of York Till,[1] which was laid down during the Illinoian glaciation, which ended about 135,000 years ago. Following that glacial period, the climate warmed, allowing new rivers to flow southerly into a lake larger and at a higher level than today's Lake Ontario, burying the York Till beneath a sequence of river delta deposits. This lake, named Lake Coleman after Coleman's death, was some twenty metres above the level of modern Lake Ontario; an ancestral Don River flowed into this lake, where it deposited a delta of muds and sands.

In a 1913 Ontario Bureau of Mines report, Coleman described what was in the lower part of these interglacial deposits, now known as the Don Formation. There were abundant fossils, including thirty-two species of trees (among them Osage orange, pawpaw and red cedar); forty-one species of shellfish (of which ten or eleven no longer live in Lake Ontario, but are found in Mississippi River waters); and many 'beetles, cyprids, etc.' in addition to the 'bone of a large bear and bones or horns of bison, of a deer like the Virginia red deer, and of a deer related to the caribou.'[2] All of these deposits were evidence of a climate warmer than at present – a climate more like that of Ohio's or Pennsylvania's today. Several years after this discovery, Coleman was moved to more poetically interpret the features of the Don Formation in his book *The Last Million Years*:

1 Till (sometimes called 'boulder clay') is a gathering of all sizes of material, from clay-sized particles to boulders, jumbled together and compacted.

2 Ontario Bureau of Mines, 1913. AR. 22, pt.1, pp. 244–245.

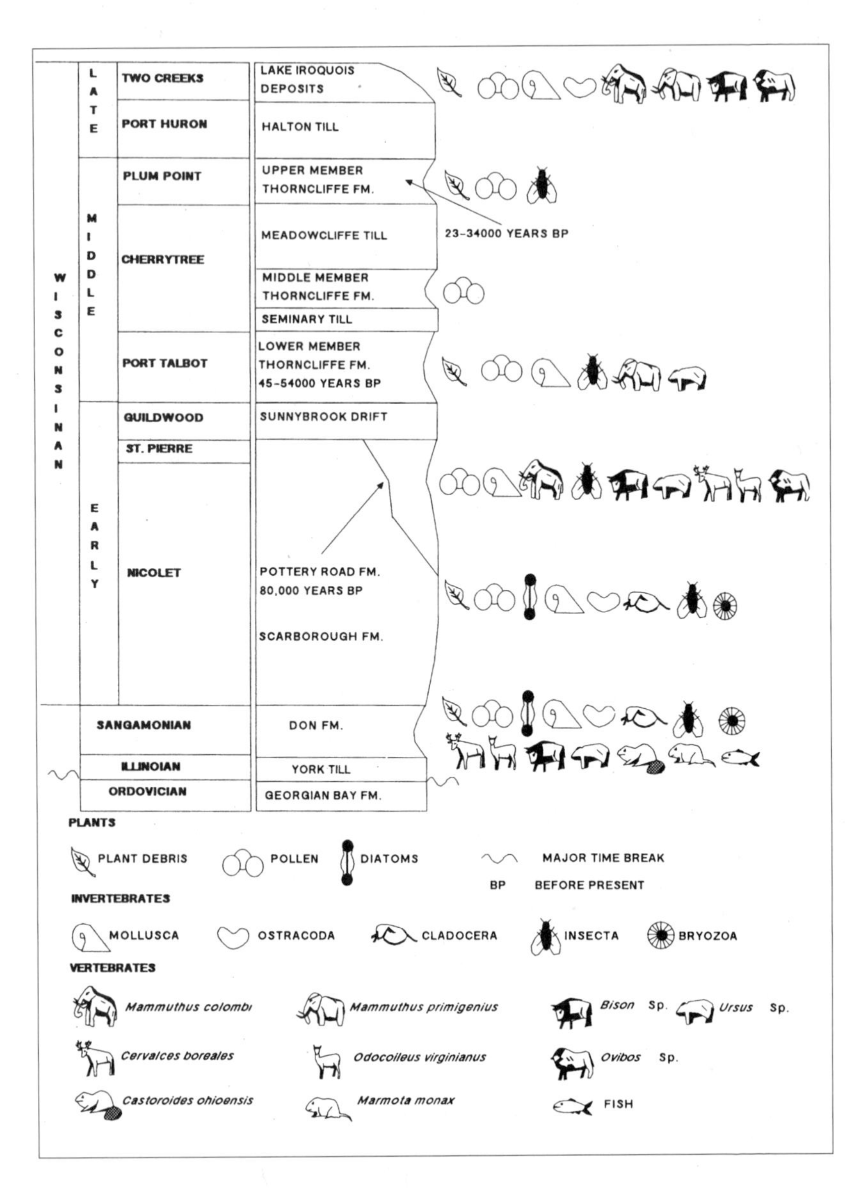


3 *Deposits at the Don Valley Brick Works, with fossil evidence shown for each deposit.* Castoroides ohioensis *is the now-extinct giant beaver.*

... one can see the ancient forest of maples and oaks and many other trees on the river shore with deer coming down to drink, bears tearing open a rotten log for its small inhabitants; and at some creek mouth the giant beaver fells a tree with a splash to feed on its branches; while openings in the forest show buffalo grazing. A thunder-storm comes up, lightning strikes a blasted tree, and fire runs along the river

> bank, stampeding the forest dwellers which rush to the water for safety – all recorded with many more features . . . in the sand and clay beds between two sheets of boulder clay . . .

Some 100,000 years ago, the last Wisconsinan glaciation began to cover the land. As the climate cooled and ice started to block the St. Lawrence outlet, the water level rose some forty-five metres higher than modern Lake Ontario to form Lake Scarborough. A new delta, the Scarborough Formation, indicative of this cooler climate, was slowly deposited on top of the warm-climate Don Formation delta. Then, beginning about 70,000 years ago, another sequence of glaciers advanced and retreated above these deltas. It is the last ice advance, some 13,000 years ago, that created the landforms that determined the drainage pattern of Toronto's streams. As glacial ice accumulates, it deforms under its own weight to 'flow' outward. When it crept westerly from the Kingston area into the Lake Ontario Basin, it pushed out of the basin. At Toronto, the ice moved from the lake basin to the northwest, whereas it moved westerly at Oakville, southwesterly at Hamilton and southerly at Niagara-on-the-Lake.

*[4] The Ridgewood Road drumlin, west of the Rouge River mouth.*

The northwesterly push of ice across Toronto moulded the surface into a sequence of linear hollows and rounded hills known as drumlins. One central Toronto drumlin is now the steep hill on Eglinton Avenue east of Bathurst Street that divides the valley of Yellow Creek and the old Belt Line Railway on its east side, from the valley of Castle Frank Brook, also known as Cedarvale Ravine, on its west side. Old Forest Hill Road runs along this drumlin's crest, providing hilltop sites for large homes. Another lofty drumlin just northwest of Eglinton Avenue and Avenue Road provides the site for a radio tower. More and higher-elevation drumlins occur to the northeast, outside central Toronto; in Scarborough, a drumlin trends northwest from Lake Ontario along Ridgewood Road, sliced by Lawrence Avenue East and the Lakeshore rail corridor [4].

But glaciations did more than just sculpt the land surface. Ice carried material and left it on the land surface as till. Throughout the GTA, rock fragments from local bedrock and the Canadian Shield have been carried and left in till as evidence of glacial movement. Canadian Shield erratics – rocks in till of a different nature than the local bedrock – found in Toronto have been used for foundations and walls, especially in

Toronto's early buildings. Glacial erratics are often found during construction; in 2002 one was uncovered near the Leaside Public Library when new gas lines were being installed, while a large boulder from north of Parry Sound was found about four metres below the surface of Robert Street during the digging of the foundation for Lansdowne Public School.[3]

After the last glacier advance of 13,000 years ago, meltwater trapped between the Niagara Escarpment and the Oak Ridges Moraine flowed southeasterly along shallow, ice-scraped depressions to create streams and ponds north and northwest of Toronto. Deposition of clay and sand in these 'Peel Ponds' created a fairly level surface originally used for farming, but which later proved to be ideal for facilities like the Lester B. Pearson International Airport. The Don and Humber rivers are two major exceptions to the southeasterly flow pattern; their paths follow surface depressions above the old pre-glacial Laurentian River channels. Humber Bay is the result of the inundation of the lower Humber by the waters of Lake Ontario, which formed a broad estuary. As late as 1893, the Humber waters were nearly four metres deep at Bloor Street. The deposition of flood debris, the result of urbanization, has filled the Humber to create the shallow river we see today.

The geological events of the past have provided the raw materials for the stone and brick to build our city. Clay deposits allowed the manufacture of brick, chimney and drainage tiles. The earliest Toronto brickyard, near the mouth of Taddle Creek, produced bricks in 1796 for the first parliament buildings. There were other brickyards in Toronto, many of them along Queen Street East, Greenwood Avenue, and Dawes and Weston roads, but the best known were those at Yorkville – Sheppard, Townsley, Nightingale – that produced bricks from the 1840s to the late 1890s and, as noted above, at the Don Valley Brick Works, which ran from 1889 to 1989.

Most of the materials used to build Toronto before World War II were found locally. Sand and gravel were obtained from the gravel bars of Lake Iroquois. The hard, silty limestone and limey sandstone lenses within the Georgian Bay Formation were used for the walls and foundations of the CAMH building on Queen Street West [5] and C'est What on Front Street. Along with erratics, stone slabs were dragged or 'hooked' up from Lake Ontario. The valley of the Humber River also

3 Some boulders in Toronto have been transported by truck, rather than ice, to celebrate events. A piece of Switzerland's Matterhorn lies near the CN Tower, Village of Yorkville Park has a 650-tonne piece of the Canadian Shield from near Gravenhurst and a Norwegian boulder marks Little Norway Park.

provided much slab stone; the Old Mill incorporates material from William Gamble's 1848 seven-storey flour mill, and is the third mill at this site that has used stone slabs from the Humber. A stone quarry was also worked throughout the late 1800s at the river below Baby Point.

When ice began its withdrawal 12,500 years ago across what was to become Toronto, the stage was set for the creation of Lake Iroquois, the largest lake yet in the Lake Ontario Basin, and the carving of Toronto's ravines. It is due to the work of A. P. Coleman and the geologists who have followed in his footsteps that we are now able to understand the dynamic changes and events of Toronto's past, and to appreciate the ancient role of water in forming and shaping the Toronto landscape.

5 *Canadian Shield boulders and local bedrock fragments form the 1851 foundation for the locally made brick wall at CAMH at 1001 Queen Street West.*

Nick Eyles

# Ravines, lagoons, cliffs and spits: The ups and downs of Lake Ontario

The bedrock basins now occupied by the Great Lakes are probably at least 2 million years old, the product of repeated phases of glacial erosion by successive ice sheets. In contrast, the lakes that fill these basins rise and fall with changing climate. The oldest water body we have direct geological evidence of is Lake Coleman, which formed some time after 80,000 years ago, when the climate in the Toronto area was subarctic and an ice sheet was beginning to expand across Canada. Lake Coleman was likely covered year-round by ice and infested by icebergs that plowed into its floor. By at least 40,000 years ago, ice had swept across all of southern Ontario, and lake waters may have survived only as subglacial lakes under the ice sheet, similar to those below the Antarctic Ice Sheet today. As the ice retreated after 12,500 years ago, another deep lake (glacial Lake Iroquois) was dammed up before abruptly draining, when ice finally left, into a much smaller lake (Lake Admiralty). The abrupt drainage of glacial Lake Iroquois gave rise to the characteristic feature of the Toronto lakeshore – its many ravines [1]. Lake Ontario came into existence about 8,000 years ago, as the basin refilled. Lake levels are still rising (albeit slowly), but much of the shoreline has now been engineered to prevent erosion and reclaim new land.

## RAVINES AS RECORDS OF CHANGES IN CLIMATE AND LAKE LEVELS

Toronto's topography is widely perceived as being flat when it is actually cut by many deep ravines that drain into Lake Ontario. Rivers and small creeks flow to the lake, imprisoned at the bottom of steep-sided, narrow valleys. Slippery side walls of muddy glacial sediments were an obstacle to early settlers and railways. Today, residences backing onto ravines are much sought after, and they are significant refuges for fauna and flora lost to urban development on the surrounding tablelands.

The ravines are of special scientific interest because they tell a remarkable story of abrupt changes in climate and lake levels during the closing stages of the last ice age. At that time, much of Canada lay entombed under the white shroud of a continental ice sheet, mammoth and bison still roamed southern Ontario and early Paleo-Indians were making their first appearance in eastern North America.

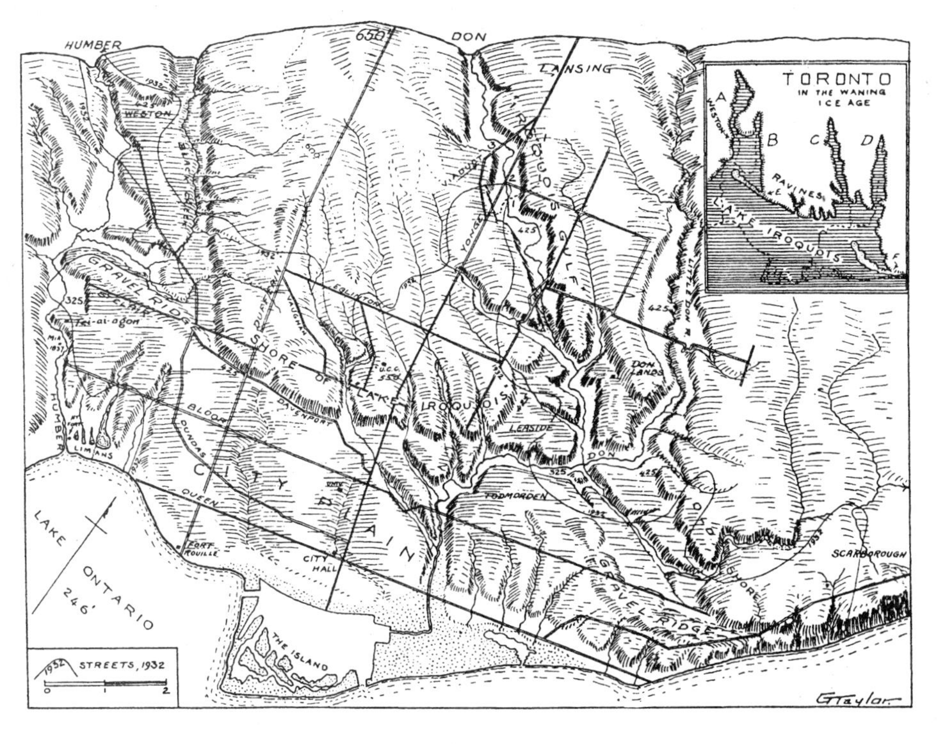


Some 12,500 years ago, the last great North American glacier – the Laurentide Ice Sheet – began leaving the Toronto area. The ice sheet's margin retreated northward back to its source areas in Labrador and Quebec, where it finally thinned and melted some 6,000 years later. Ponded in front of the ice margin was glacial Lake Iroquois, which flooded much of the Toronto area. This very deep lake drained to the Atlantic Ocean not via the still-ice-blocked St. Lawrence Valley, but through the ice-free Hudson River Valley in upper New York State. At this time, downtown Toronto would have been under some sixty metres of water (the height of a twenty-storey building), in which drifting icebergs floated after having calved from the ice front [2].

Glacial Lake Iroquois cut a prominent shore bluff and beaches that can be mapped all the way around Lake Ontario and that record the previous high-water mark. Casa Loma, one of Toronto's best-known landmarks, sits prominently atop the Iroquois bluff, gazing out over the city below; early settlers used cableways to ascend and descend the steep slope.[1] At the foot of the Iroquois bluff there are beach gravels and sands

1 *Griffith Taylor's 1936 sketch of the Toronto area showing the principal ravines and the shoreline of glacial Lake Iroquois.*

1 The lake was so named in reference to the use of the beach as a track by the Iroquois before the arrival of Europeans.

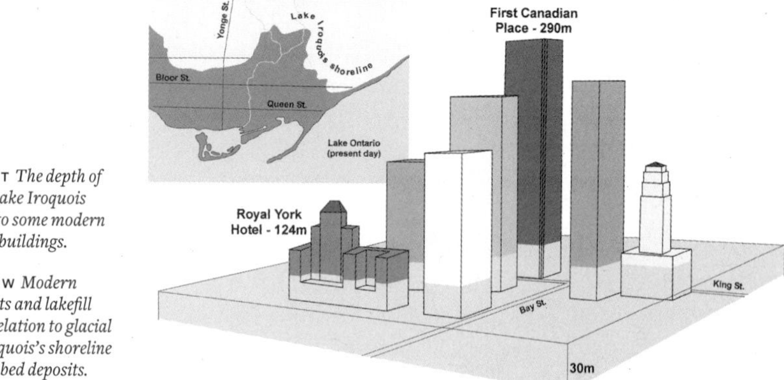


**2** RIGHT *The depth of glacial Lake Iroquois relative to some modern Toronto buildings.*

**3** BELOW *Modern gravel pits and lakefill sites in relation to glacial Lake Iroquois's shoreline and lake bed deposits.*

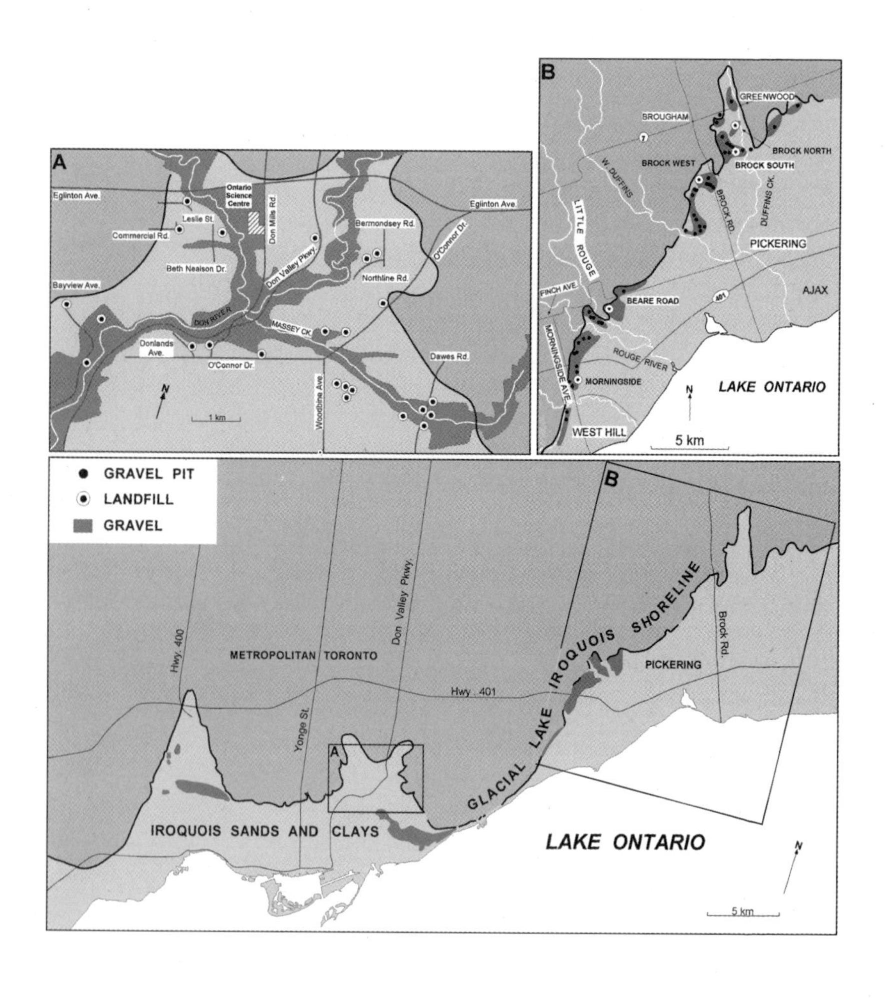

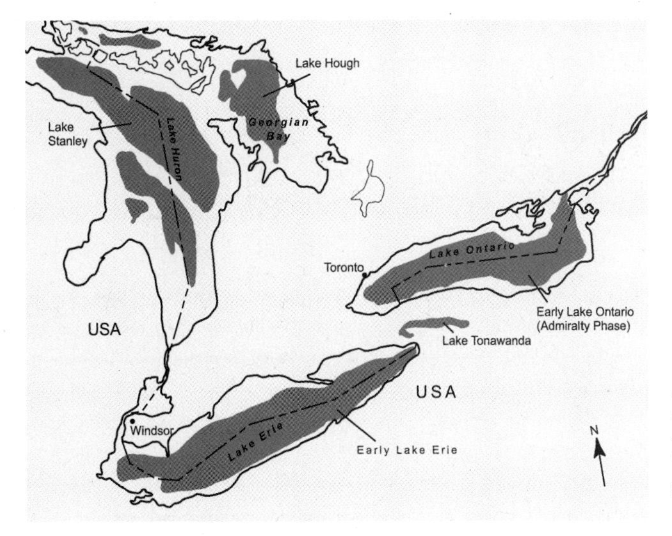


[4] *Lakes Ontario, Erie and Huron today and 10,000 years ago. In its Admiralty phase, Lake Ontario's shoreline was five kilometres south of Toronto.*

that, once quarried, left large pits that were later filled with waste. The extent of the Iroquois shoreline around Toronto can be mapped with reference to the many hills created by landfilling [3].

As the ice front slowly melted back northeastward, the outlet to the St. Lawrence Valley, roughly where Kingston is today, suddenly opened. This opening had the effect of a plug being removed from a bathtub, causing Lake Iroquois to abruptly drain about 12,200 years ago. At this time, however, southern Ontario and Quebec were tilted eastward under the great weight of the thick ice sheet (an effect called 'glacioisostatic depression'), and sea level was some 100 metres below where it is today because enormous amounts of fresh water were still locked up in the ice sheet. In combination, the eastward tilt and low sea level meant that the early Great Lakes drained almost entirely, leaving small remnant lakes: imagine a tilted bathtub whose waters flow unimpeded from the plughole. Rivers that had drained to glacial Lake Iroquois now flowed out many tens of kilometres to the distant shoreline of a much smaller early Lake Ontario – the Admiralty phase of Lake Ontario [4]. In flowing to a much lower lake level, the rivers were 'reactivated,' and energetically cut down into the glacial sediments left by the ice sheet, excavating a huge amount of sediment. This is the origin of Toronto's many ravines: they are an exceptional record of the low lake levels that existed many thousands of years ago, and the deep erosion caused by rivers.

## THE LAST 8,000 YEARS

Low lake levels persisted in the Great Lake basins until at least 8,000 years ago, when the east-tilted crust began to rapidly recover in a process geologists call 'postglacial rebound.' The raised sill at the east end of Lake Ontario resulted in slowly rising water levels in the lake basin. This eastward-increasing crustal rebound is still taking place today, only much more slowly. Lake depths are increasing at the western end of Lake Ontario at Hamilton as the Kingston area slowly continues to rise. In a few thousand years, this readjustment process will end (and the next glaciation may start!).

A cliff line recently found some five kilometres offshore of the modern shoreline of Lake Ontario marks the early postglacial Admiralty phase. This feature is easily recognized on bathymetric charts (the submerged equivalent of an above-water topographic map) and is known locally as the 'Toronto Scarp.' It is a favourite spot for salmon that like deep, cold waters. By allowing intake pipes to reach icy cold water in a relatively short distance, it is also a feature that has made Enwave's deep-lake water-cooling project possible.

## RISING LAKE LEVELS AND DROWNED RAVINES

Rising lake levels in Lake Ontario over the past 8,000 years have bequeathed another landscape feature now found at the lakeward ends of the ravines. The mouths of the ravines are being slowly drowned, trapping wetlands and lagoons behind sandy spits. Hamilton Harbour, Cootes Paradise, Grenadier Pond and the many lagoons that dot the shoreline of Lake Ontario from Niagara-on-the-Lake to Cobourg record the landward movement (or 'transgression') of the shoreline over the past few thousands of years [5]. This also explains why there were large wetlands at the mouths of rivers such as the Humber and Don.

There is some evidence of an accelerated rise in the level of Lake Ontario some 4,000 years ago, but the origins of this rise are not clear. This time does broadly coincide with a short-lived phase of global cooling geologists call the 'Neoglaciation' because it saw the regrowth of glaciers in the Canadian Rockies and elsewhere and may have been a time of wetter climate in the Great Lakes. On the other hand, some have suggested that the slow rise in the crust to the northeast of the Great Lakes had the effect of diverting waters to the southwest back toward the lakes.

5 *The spit-and-lagoon formation at Frenchman's Bay, just east of Toronto.*

## RISING LAKE LEVELS AND COASTAL EROSION

The prominent cliffs seen at the Scarborough Bluffs are the result of the shoreline moving inland by eroding into soft glacial sediments. The cliffs at Scarborough once lay some five kilometres offshore, and have retreated landward over the past 8,000 years as the level of Lake Ontario has slowly risen. Historically, this has occurred at a rate of about one metre per year, but it has slowed in the last twenty years. Over the past 8,000 years, enormous volumes of sand, eroded from the bluffs and moved westward by strong lakeshore currents as a consequence of the shoreline being extensively eroded, have fed the growth of the Toronto Islands, which are, in essence, large spits. Today, these bold cliffs have been straitjacketed by engineers, greatly reducing erosion and the supply of sand. The prominent marina at Bluffers Park, where Brimley Road meets the lake, traps sand on its up-current (east) side but starves beaches down-current. The desire to protect private property from erosion has also armoured much of the coastline under a layer of asphalt and rip-rap, except for a small portion at Dutch Church, the most dramatic geological feature of the bluffs. These towering cliffs have long been the trademark of Scarborough, and explain why the area was so named by Elizabeth Simcoe in 1793, in recognition of similar cliffs in Scarborough, England.

**6** RIGHT *The Scarborough Bluffs at Sylvan Park, 1999. Much of the shoreline has been armoured, and precious outcrops of sediments recording past climates have been lost to study.*

**7** BELOW *Geological layers at the Don Valley Brick Works, 1985. The bedrock quarry has since been filled in. The York Till records the penultimate glaciation (the Illinoian), and the Don Beds the warm interglacial period that followed some time around 110,000 years ago. Younger overlying sediments record the last glaciation (the Wisconsinan). The flat terraced top of the outcrop was cut by glacial Lake Iroquois when the last ice sheet left the Toronto area shortly before 12,000 years ago.*

Unfortunately, these cliffs will disappear as the temptation to build a waterfront trail at their foot becomes too much for the Toronto and Region Conservation Authority (TRCA). The cliffs will slowly disintegrate and become vegetated, as has happened to the east and west. This is potentially a great loss, not only of a historic landmark, but of an internationally significant record of climate change [6].

Man-made changes have been no less dramatic elsewhere. Wetlands have been drained and lagoons infilled across Toronto. The size and shallowness of Ashbridge's Bay at the mouth of the Don River have made it desirable and relatively easy to reclaim land in the area now known as the Port Lands.

Across Toronto, the glacial Lake Iroquois shore bluff played a key role in acting as a source of springs. These fed the numerous creeks that flowed south across the downtown and midtown areas. Sadly, these have been crossed by roads and filled with waste to make new land for development. Their rivers have been straightened and imprisoned within pipes and channels, and their gravel floors mined to make concrete for the growing city.

The story of human intervention along the creeks and shorelines is not entirely bad, however. Clay excavations along the banks of the Don River for brick-making in the 1890s revealed ancient sediments from a warm climate episode (an interglacial) older than the last glaciation. The Don Valley Brick Works is now known to be among the best preserved interglacial records found anywhere in northern North America [7]. Most famously, it preserves the remains (mostly teeth) of the extinct giant beaver. This animal is a fitting symbol of the remarkable ups and downs of lake levels in the Great Lakes.

Ronald F. Williamson & Robert I. MacDonald

# A resource like no other: Understanding the 11,000-year relationship between people and water

The Lake Ontario shoreline and its rich estuaries and river systems have been home to people for at least 11,000 years. Archaeological sites are almost all we have to tell the stories of those early settlements, whose physical remnants are very fragile. Finding ways to protect those remnants has become especially important in the last half-century of urban sprawl and expansion. The rate of loss of archaeological sites during that period is staggering – it is possible that hundreds of sites were destroyed in Toronto alone, many of which would have contributed meaningfully to our understanding of the past.

Until the environmental protection legislation of the 1970s and 1980s, municipalities had few tools to combat archaeological site destruction. Armed now with unequivocal instructions from the nation's supreme courts that outline the need for consultation with the Aboriginal communities whose legacy these sites represent, and that have given shape to a clear legislative expression of the provincial interest in the archaeological record, municipal planners must now require the identification of archaeological sites on land poised for development before the backhoes touch the ground.

The primary means by which these resources can be protected is through the planning approvals process. Archaeological master-planning is the latest response to an old problem: how to deal with evidence of the past that is, for the most part, not visible because it is buried underground (or under water). By providing an inventory and evaluation of known archaeological sites, as well as a model that identifies where other sites might be situated, archaeological master plans are the new tool of choice for planners, helping them determine when to call for archaeological assessments in advance of development. Toronto is undertaking one of the most ambitious and comprehensive of these plans in North America.

Almost 150 Aboriginal archaeological sites have been found in Toronto and registered in the provincial database since 1974. These sites date from the earliest period of human occupation in the region through to the nineteenth and early twentieth centuries, and represent a wide range of settlement types, from places where pre-contact hunters lost their spears thousands of years ago to major 500-year-old farming villages that housed over a thousand people and were surrounded by hundreds of hectares of cornfields.

To predict how additional but undocumented Aboriginal archaeological sites are distributed across the Toronto landscape, consulting firm Archaeological Services Inc. designed an archaeological potential model based primarily on the distance to various forms of potable water. Soil drainage and texture characteristics (especially for agriculturalists) and slope attributes were also considered. The universality of the need for potable water makes its examination a logical point of departure for most predictive modelling exercises. Aboriginal elders and oral histories tell us that the only exceptions usually relate to special places in the landscape where people communicated with the spiritual world, such as rock outcrops or caves for vision quests or burials. The fact that the average distance to water for the eighteen burial sites in Toronto is only 185 metres suggests that people may have believed that those souls that still resided with their ancestors' bones also required water for sustenance. Many Aboriginal groups believed (and still do) that people have two souls, one that goes to the sky world at death and one that stays with the person's remains, which accounts for why such places are considered sacred and should not be disturbed.

While water is arguably the most fundamental resource humans require, archaeologists recognize that the natural landscape of southern Ontario has not remained the same during the span of human occupation. We need to understand where water was during any particular period in the past. Fluctuations in the water levels of the Great Lakes basins had profound effects on early pre-contact settlement and subsistence patterns, alternately opening up and then covering vast areas of land, affecting the survival and present accessibility of sites. This poses a particular problem for finding sites from the earliest occupation of the city, when the Toronto shore of Lake Ontario was several kilometres south of its current location.

Although the character of Toronto's rivers and their tributaries has been severely obscured by land development, we know from historical accounts that, just 200 years ago, they were far different from today. The Don River, for example, was navigable upstream by boat for three or four kilometres from Lake Ontario and was bordered along its lower reaches and around its mouth by wetlands. In his 1873 history of Toronto, *Toronto of Old*, noted cleric and scholar Henry Scadding described the contiguous marshes through which the Don flowed as 'one thicket of wild willow, alder, and other aquatic shrubbery,' including witch hazel, dogwood, highbush cranberry, wild grape, blue iris, reeds and cattails. He refers as well to an island near the mouth of Castle Frank Brook where wild rice grew plentifully. His account aptly describes the reasons that archaeological sites are close to rivers – they provide potable water, a means of transportation by canoe and, when surrounded by marsh, a veritable supermarket of plant and animal foods, as well as raw resources for tools and clothing.

Understanding the evolving natural environment, and especially water, is crucial for archaeologists attempting to predict where past peoples lived throughout the millennia. The investigations of known archaeological sites have provided rare and valuable insights into people's past relationships with the water in their environments. The following sections highlight some of those insights, from the time of the very first occupants of the region and their use of ancient glacial shorelines to the long-distance travel and trade along the major regional rivers that characterized the ancestral Huron and Seneca use of the region. Throughout the millennia, the trails alongside the Rouge, Don and Humber rivers functioned as the pre-contact equivalents of Highway 400 in the linkage of the lower and upper Great Lakes.

## THE VIEW FROM THE IROQUOIS BLUFF (9000 BC–7000 BC)

Small bands of nomadic hunters moved to the southern Ontario region soon after the continental glacier retreated. They often settled on ancient shorelines where they could spot large game such as caribou, mastodon or mammoth in what was then a tundra-like environment similar to that found today in the subarctic region of Canada [1]. The Lake Iroquois strandline above Davenport Road is one such relict shore, although it was located well inland by the time people first moved into the Toronto area. While only a few cultural traces of this period have been found within the city, dozens of 10,000- to 11,000-year-old spearheads have been found along this ancient shoreline in neighbouring municipalities. Unfortunately, some of the largest campsites dating to this

period would undoubtedly have been situated adjacent to the Lake Admiralty shoreline or the estuaries that drained into that lake, which are now situated some five kilometres into Lake Ontario.

*[1] The earliest occupants of Toronto may have camped on a similar glacial shoreline to spot caribou or other large game in the distance.*

Truly spectacular evidence of those submerged occupations may have surfaced in 1908 during waterworks-related tunnelling in Toronto Bay, to the east of Hanlan's Point, at a depth of twenty metres below water level. This evidence consisted of over 100 human (possibly moccasined) footprints in clay that were likely laid down during the Wisconsinan glaciation. Although not professionally investigated at the time due to a fear of delay in construction, the on-site inspectors clearly described in newspaper interviews both child and adult footprints, perhaps representing a family heading northward from their camp on the Admiralty lakeshore to what is now downtown Toronto some 11,000 years ago.

HUNTING AND FISHING ON TORONTO'S RIVERS (7000 BC – AD 500) For the next several thousand years, the landscape in which people lived continued to change. The water levels in the Great Lakes lowered and broadleaf forests expanded, necessitating changes in the hunting, fishing and gathering strategies. This in turn brought about new weapon

2 *Two sides of a 4,000-year-old spear head found on a site near James Gardens on the Lake Iroquois shoreline in the Humber Valley.*

and tool technologies including, eventually, the bow and arrow.

By 4,000 to 5,000 years ago, small bands of related families were settling into familiar hunting territories. They spent the spring and summer in large settlements located near river mouths, fall in small camps in the forested interior so they could harvest nuts and hunt deer, and winters in even smaller camps.

While the earliest of the lakeshore and estuary sites are now either submerged or buried under modern lakefill, a number of the interior sites have survived. The distribution of the thirty-five sites of this period found in Toronto is divided equally between those situated on major rivers and creeks and those situated along small tributaries. A handful of sites are also adjacent to wetlands, and at least two sites are known along the Iroquois shoreline adjacent to the Humber River, including a recently documented 4,000-year-old camp in James Gardens [2]. Other nearby sites are located along the middle reaches of the Humber and Don rivers. The collections from sites in the Eglinton Flats area, for example, contain projectile points that date to the period from 1,000 to 400 BC.

The harvesting calendar of this period would not have been complete without the seasonal catches of fish from Lake Ontario and the lower reaches of the major rivers. While people fished throughout the year, they would have had much larger catches during the spawning runs in spring and fall, as evidenced by the presence of at least seven known sites along the lower Humber, Don and Rouge rivers. Whether situated in forest interiors next to small tributaries or wetlands or adjacent to major rivers, the sites of this period are rarely situated more than 200 metres away from water.

## THE ANCESTRAL HURON: FIRST FARMERS (AD 500 – 1600)

The introduction of corn into southern Ontario about 1,600 years ago brought a commitment to producing food through agriculture. The seasonal rounds of the previous millennia were abandoned in favour of growing all food in one place. In the first centuries after the introduction of corn, life involved clearing land first on flood plains and, shortly thereafter, on the tableland surrounding small base settlements. People tended their fields from these settlements while sending out hunting, fishing and gathering parties to satellite camps to harvest other naturally occurring resources, thereby reducing

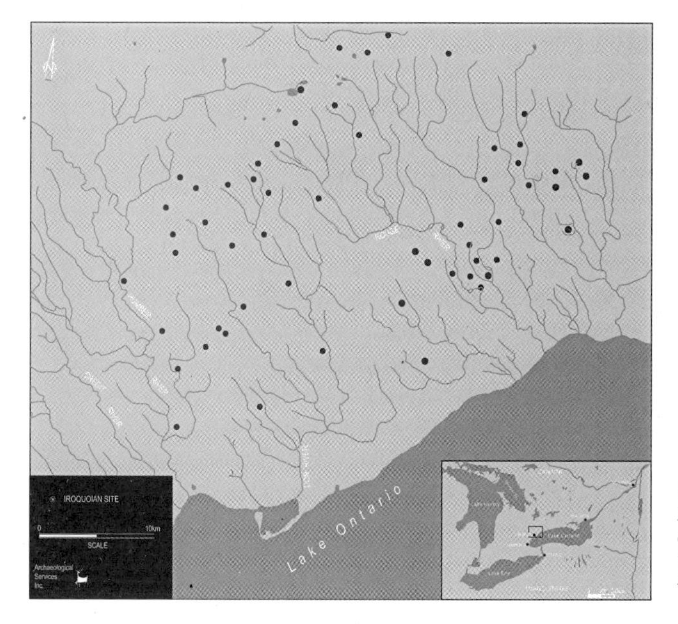


**3** *Locations of ancestral Huron and Petun sites along the rivers draining into the north shore of Lake Ontario, 1350–1550.*

the risk posed by crop failure. By the thirteenth century, these base settlements had grown to one hectare or more, and squash, beans, tobacco and sunflower were grown in addition to corn.

The first agriculturalists who lived along the central north shore of Lake Ontario in the Toronto region were the ancestors of the Huron and Petun [**3**]. By the beginning of the fourteenth century, there were a number of contemporaneous communities occupying the Humber, Don, Highland and Rouge drainage systems, some of which probably still remain to be documented. While these sites were usually around two hectares in size, by the early sixteenth century, they had grown to four to five hectares, containing at any one time more than fifty house structures occupied by 1,500 to 2,000 people [**4**]. The resource needs of these early towns included 60,000 trees for house and palisade construction, probably taken largely from nearby cedar swamps. Given the requirement of one pound of corn per day per person, hundreds of hectares of land would also have been cleared around the villages for cornfields. As no irrigation or fertilization methods existed, the ever-decreasing soil fertility would have resulted in field expansions to the point that cabins were necessary in the fields to eliminate the several-kilometre daily walk back and forth from the main village to the remote fields.

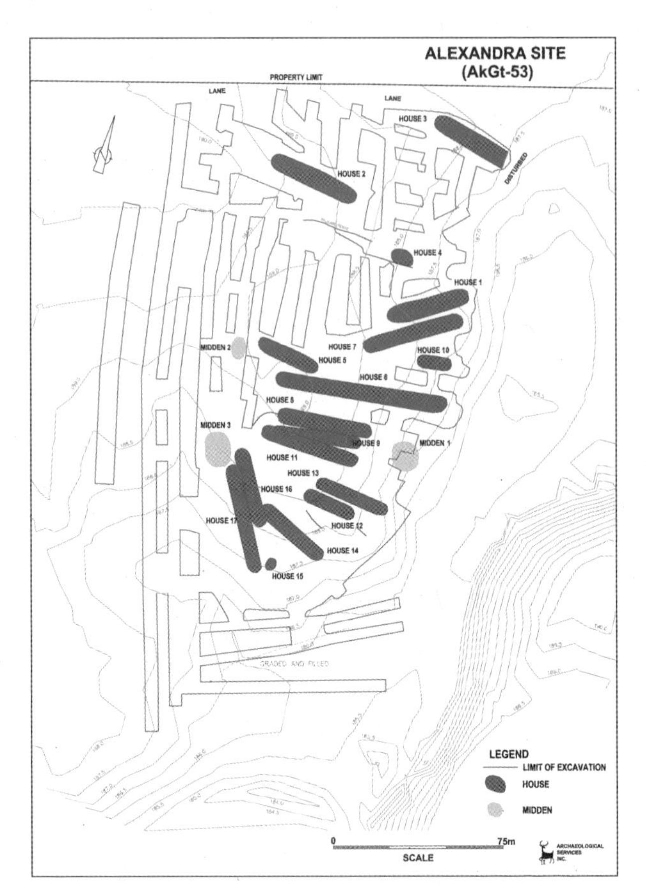


4 *Plan of the Alexandra site, a fourteenth-century ancestral Huron settlement on Highland Creek in northeast Toronto. Excavated in 2001–02, the site yielded evidence of sixteen longhouses and several refuse deposits in which over 19,000 artifacts were found.*

There have also been a number of discoveries of ancestral Huron ossuaries in the city.[1] One such late-thirteenth-century site was documented on a small tributary of the Don River just south of Highway 401. Test excavations of the village, known as the Moatfield village, and its associated ossuary, yielded large numbers of fish remains, along with turtles, waterfowl and a variety of both land and water mammals. These finds show how important the resources of the lower Don River were for the site inhabitants. In fact, in addition to American eel, one group of fish comprising Atlantic salmon, lake whitefish and lake trout played a particularly significant role in the diet of the site inhabitants, providing nutritional balance to a maize-dominated regimen. Along with pickerel, these species, all high in the food chain, left a chemical signature in the bones of the village occupants in the form of high nitrogen isotope values.

1 Ossuaries are pits three metres in diameter in which the remains of hundreds of people who had died during the tenure of a village were placed and then commingled.

[5] *A 1688 map by Pierre Raffeix shows the location of Iroquois villages along the north shore of Lake Ontario, including Teiaiagon on the Humber and Ganatsekwyagon on the Rouge. Portage routes to 'Lac Tarontho' (Lake Simcoe) are traced. This is the first European map to show the Don River. North is at the bottom.*

Similar data from other fifteenth-century communities in the region situated more than fifteen kilometres from the lakeshore suggest that deer and other mammals were eaten to achieve that balance. Scientists have also been able to examine the oxygen and strontium isotopes in people's skeletal remains to determine where those people were from, since the chemical signatures of those elements are typically acquired through local foods and drinking water.

Most if not all of the communities along Lake Ontario's north shore had moved northward by about 1600, joining with other groups in present-day Simcoe County (between Lake Simcoe and Georgian Bay) to form the Petun and Huron confederacies documented by the first European visitors to the area. By the mid-seventeenth century, those populations had been dispersed due to European disease, resulting in a loss of over half of their populations, and war with the Five Nations Iroquois from south of Lake Ontario, who were traditional enemies of the Huron and in constant conflict with them over beaver hunting grounds and access to the lucrative European fur trade.

THE IROQUOIS AND MISSISSAUGAS: MID-SEVENTEENTH-CENTURY NEWCOMERS (AD 1650–1700) The Five Nations Iroquois established a series of settlements at strategic locations along the trade routes inland from the north shore of Lake Ontario, including two villages in Toronto [5]. These new settlements were established by the Seneca near the mouths of the Humber and Rouge rivers, two branches of the

Toronto Carrying Place – the canoe-and-portage route that linked Lake Ontario to Georgian Bay and the upper Great Lakes through Lake Simcoe. By the 1690s, both villages were home to Mississaugas, the Seneca having abandoned the villages to return to their homeland to defend it from the French. Though these villages were similar to earlier Huron settlements along the north shore in that they were surrounded by cornfields, the inhabitants of both villages also depended on the rich salmon fisheries near the mouths of the rivers.

The Seneca settlement on the Humber River is called 'Baby Point' or 'Teiaiagon' (an Iroquois word meaning 'cross the river') and was situated on the level summit and slopes of a large promontory overlooking the main channel of the Humber. The Récollet missionary and explorer Father Louis Hennepin spent three weeks at the settlement in the late autumn of 1678, and French explorer René-Robert Cavelier de La Salle camped at the site in the summers of 1680 and 1681. While hundreds of graves were documented on the site in the early twentieth century, the bodies of two Seneca women were recently discovered, both with ornamental combs, one of which is carved to depict morphing animal figures including Mishipeshu, a powerful underwater dragon-like being [6].

The other Seneca village in eastern Toronto was called 'Ganatsekwyagon' ('among the birch trees'), situated near the mouth of the Rouge River. The first European use of the site was as a mission established by the Sulpician Fathers from 1669 to 1671 under François d'Urfé.

The Senecas, the Mississaugas and the earliest Europeans along the present-day Toronto waterfront were probably there because of the area's strategic importance for accessing and controlling long-established regional economic networks. All these occupations occurred on or near the Lake Ontario shoreline between the Rouge and Humber rivers, at sites that had both natural landfalls for Great Lakes traffic and convenient access, by means of the various waterways and overland trails, into the hinterlands. The origin of the place name Toronto, likely a Mohawk word, means 'trees in water,' which may refer to the fish weirs in the narrows of Lake Simcoe, the first major body of water on the Carrying Place. In the end, the first European settlement of Toronto was simply a continuation of patterns that had been in place for thousands of years – patterns that were centred on water.

6 *A late-seventeenth-century Seneca moose-antler comb depicts a number of significant iconographic figures (including Mishipeshu, a bear and a human) that morph into one another. Etched lines depict Algonkian-Iroquoian symbols such as the power lines that emanate from the animal figures.*

The understanding of the past derived from ongoing research for Toronto's archaeological master plan, with its focus on people's long-standing relationship to water, should remind us of the importance of properly conserving this most fundamental human resource for our own future. Toronto's archaeological master plan has provided the City of Toronto with the tools to manage, preserve and interpret its 11,000-year-old archaeological record. Many ancient settlements have survived in the green spaces of Toronto, and we can now look forward to their conservation and interpretation, a situation that was unimaginable only a few decades ago. New initiatives that will involve First Nations in the management and interpretation of their past represents an equally significant advance that provides us with a partnership for stewardship of the past, and of the present and future as well.

Chris Hardwicke
& Wayne Reeves

# Shapeshifters: Toronto's changing watersheds, streams and shorelines

*'The Valley of the Don, a winding stream that flows on the east of the city, offers the prettiest bit of scenery in the neighbourhood.'* (Canadian Illustrated News, *June 10, 1871*)

Mutability is part of the fluid character of water. Some dynamics are natural, like the meander of a stream across its flood plain, its fluctuations in volume across the year, the cyclical rise and fall of lake levels, the building up and breaking down of sandspits, dunes and beaches. Over the past 200 years, however, humans have played a more critical role than nature in reshaping the aqueous landscape in Toronto. A good way to understand the character and impact of change is to take some big-picture snapshots of Toronto's watersheds, streams and shorelines, sketching out both the current shape of these features and a sense of how Euro-Canadians first knew them. It's primarily about providing extended captions for lines on maps – examining the structure and form of Toronto's surface water, rather than its quantity or quality.

**1. LAKE ONTARIO SHORELINES BETWEEN THE HUMBER RIVER AND VICTORIA PARK AVENUE, 1834–2008**

Toronto's waterfront has been a most movable entity. Since the mid-1850s, repeated waves of lake-filling have made land for railway entrances, industry, port functions and recreation across Toronto. Many agencies were involved at different times in this transformation, including the city government, the Toronto Harbour Commission (now the Port Authority), the provincial government and the TRCA. Notable moves evident at this scale include the filling in of Ashbridge's Bay to create the Port Lands, which mainly took place between 1914 and 1930, and the creation of the five-kilometre-long Leslie Spit, beginning in 1959. Toronto Island has also grown (though Gibraltar Point has seen rapid erosion in recent years), and new purpose-built lakefill parks – Ontario Place and Ashbridge's Bay Park – were added in the 1970s.

Perhaps the one constant in all this work is the progressive hardening of the waterfront edge – a process that also holds true for most surviving watercourses in Toronto. The imposition of engineered geometries over the softer profiles of nature is well-illustrated by the Ship Channel, developed by the Harbour Commission in the 1910s to access the interior of the Port Lands. This concrete-walled feature, perhaps best appreciated from the Cherry Street bascule bridge, now extends over 2,800 metres – the distance from Yonge Street to Strachan Avenue.

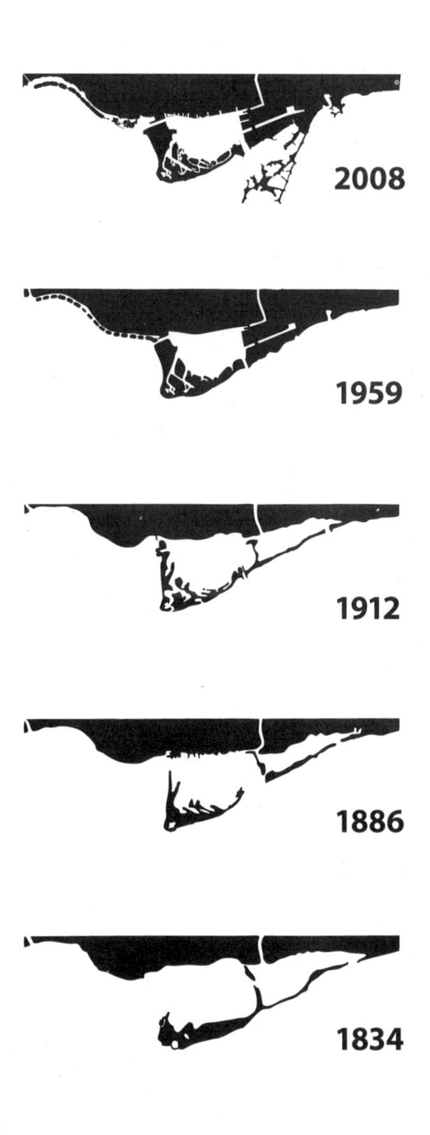

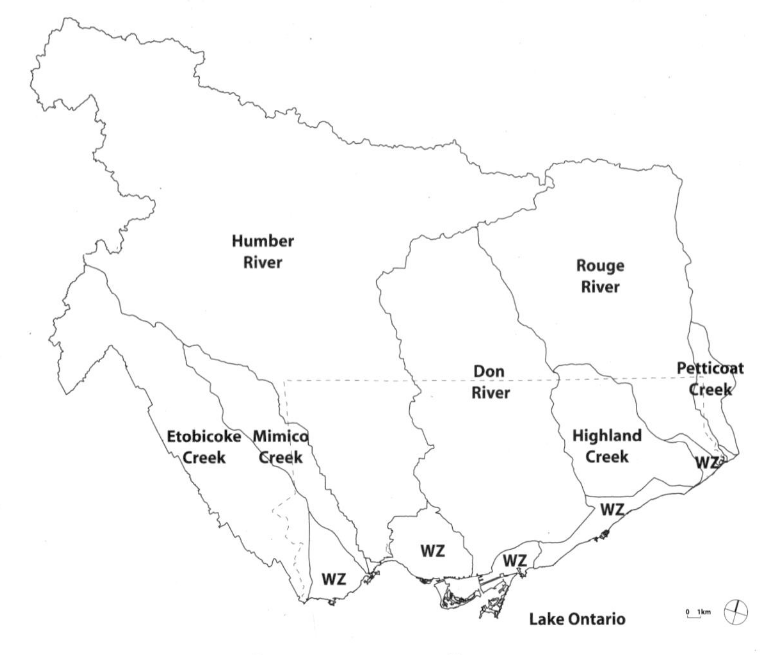


**2. TORONTO'S WATERSHEDS, 2008**

The *Canadian Oxford Dictionary* defines a watershed as 'the area drained by a single lake or river and its tributaries.' Seven named watersheds currently pass through the City of Toronto. The Etobicoke, the Mimico, the Highland and the Petticoat are the smaller drainages, rising at relatively low elevations on the flattish South Slope and Peel Plain below the Oak Ridges Moraine. The largest watersheds – the Humber, the Don and the Rouge – have their headwaters higher up on the rugged moraine,[1] the crest of which marks a north/south divide for Toronto's surface water.[2]

Toronto's watershed complexities are highlighted in data supplied by the TRCA.[3] The Don watershed makes up one-third of Toronto's land area, and nearly 60 percent of the watershed lies within the city. The Don has been front and centre in Toronto's mind since the British planted the town site in 1793. And given how the Don sprawls across the centre of the city, stretching from Keele almost to Brimley Road and taking in most of Yonge Street, we can justify viewing the Don as 'Toronto's river.' Highland Creek is, however, another worthy candidate for this title. It may occupy only 15 percent of Toronto's landmass, but 95 percent of the Highland's watershed lies within the city. The Humber occupies the middle ground. It's easily the largest watershed in the Toronto region – at

1 The Humber's headwaters are actually shared between the Oak Ridges Moraine and the even more impressive Niagara Escarpment.

2 *Surface* needs emphasis because groundwater is a much more mysterious entity. One hydrogeological study found that groundwater coursing through the face of the Scarborough Bluffs had its source in Barrie, on the far side of the moraine.

3 What (major) watershed do you live in? The answer is on a detailed Toronto region map in the 'Resources' section at www.torontorap.ca.

| *Watershed* | *Area of watershed in City of Toronto (ha)* | *Watershed as percentage of City land area* | *Total watershed area (ha)* | *Percentage of City land in watershed* |
|---|---|---|---|---|
| Don River | 20,632.6 | 32.5 | 35,806.0 | 57.6 |
| Etobicoke Creek | 1,478.6 | 2.3 | 21,164.8 | 7.0 |
| Highland Creek | 9,614.0 | 15.1 | 10,157.8 | 94.6 |
| Humber River | 13,731.9 | 21.6 | 91,077.8 | 15.1 |
| Mimico Creek | 2,900.5 | 4.6 | 7,709.1 | 37.6 |
| Petticoat Creek | 240.3 | 0.4 | 2,682.2 | 9.0 |
| Rouge River | 3,395.5 | 5.3 | 33,288.8 | 10.2 |
| Waterfront Zone (wz) | 11,587.9 | 18.2 | 11,587.9 | 100.0 |
| TOTAL | 63,581.3 | 100.0 | 213,474.4 | 29.8 |

91,000 hectares, more than two-and-a-half times as big as the Don – but occupies 'only' 22 percent of Toronto's area.[4] The Etobicoke, the Mimico and the Rouge each take in between 2 and 5 percent of the city. The Petticoat Creek watershed is so far out on Toronto's northeastern periphery, it barely registers in our civic consciousness. It occupies only 240 hectares, or 0.4 percent, of Toronto's land area, and lies entirely, and rather confusingly, within Rouge Park. The stream itself flows through Markham and less than a kilometre of Toronto before emptying into Lake Ontario in Pickering.

To muddy the waters even further, the TRCA adds a category called the 'waterfront zone.' The WZ comprises six discrete tracts of land along the lakeshore that span a remarkable 18 percent of Toronto's land mass. These tracts don't drain into any of the big rivers, and they now have little, if any, surface water of their own. These are the historical drainages of Toronto's lost rivers. The zone between the Etobicoke and Mimico creeks includes North Creek and Jackson Creek. It's not clear which minute streams drained the zone between Mimico Creek and the Humber River. The zone between the Humber and the Don rivers takes in Taddle Creek, Garrison Creek, Russell Creek and tinier streams that flowed during the early days of the Town of York. Toronto Island is its own zone, once full of many lagoons and ponds, but too sandy for running water. The fifth zone extends from the Don River to Highland Creek, embracing creeks that once flowed into Ashbridge's Bay, plus the deep gully drainages along the Scarborough Bluffs.[5] The sixth zone, between the Highland and the Rouge, continues to carry fragments of Adams Creek.

4 The Humber River is unique in that, since 1999, it's been Toronto's only designated Canadian Heritage River. The Canadian Heritage River System was established in 1984 by the federal, provincial and territorial governments to conserve and protect the best examples of Canada's river heritage, to give them national recognition and to encourage the public to enjoy and appreciate them.

5 Water still flows for most of the year in Gates Gully, the largest ravine along the bluffs, which is located near the foot of Bellamy Road. In the nineteenth century, this ravine was the principal means of accessing the Lake Ontario shoreline from Scarborough's tablelands.

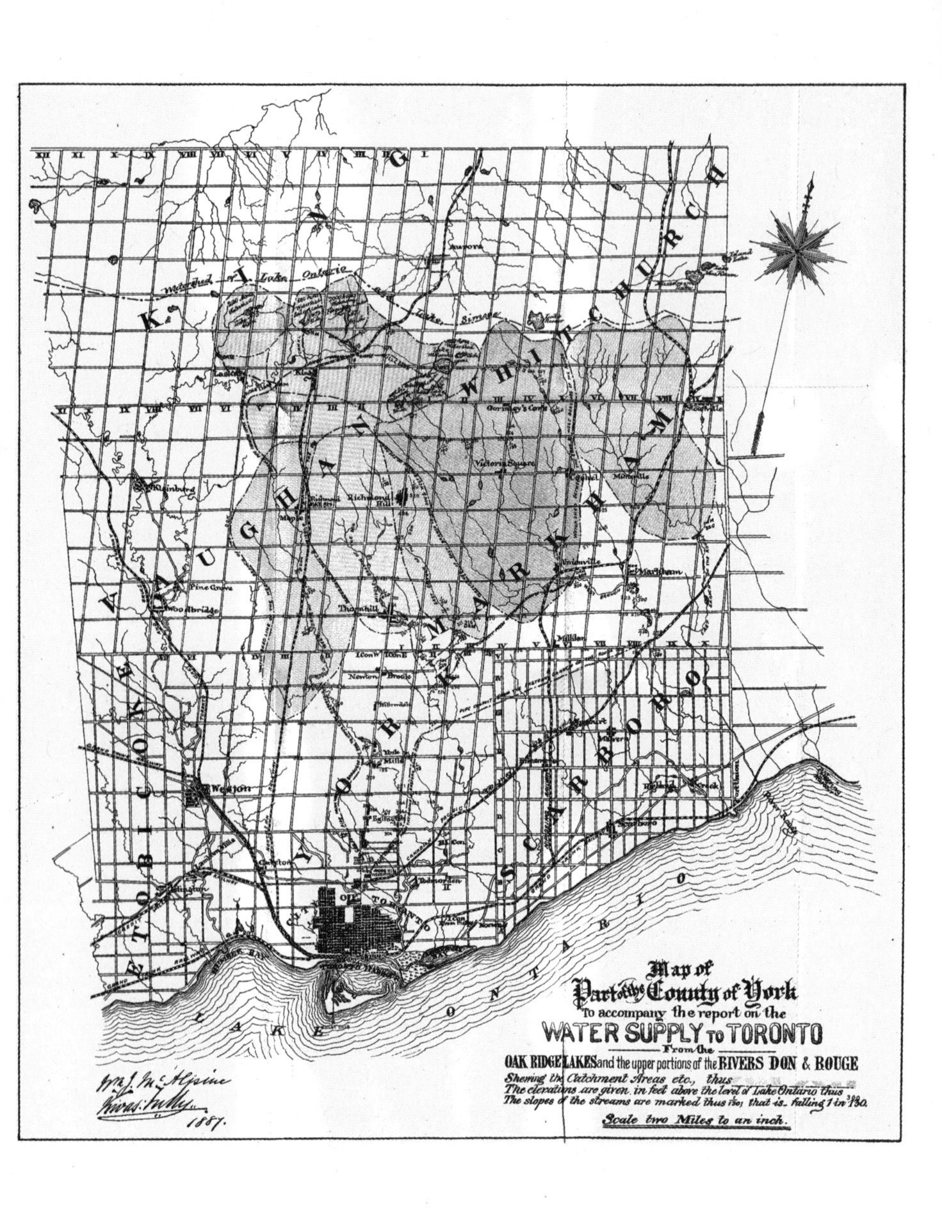
Map of
Part of the County of York
To accompany the report on the
WATER SUPPLY TO TORONTO
From the
OAK RIDGE LAKES and the upper portions of the RIVERS DON & ROUGE
Shewing the Catchment Areas etc., thus
The elevations are given in feet above the level of Lake Ontario thus
The slopes of the streams are marked thus; that is falling 1 in 380.
Scale two Miles to an inch.
1887.
KING
WHITCHURCH
VAUGHAN
MARKHAM
ETOBICOKE
YORK
SCARBORO
LAKE ONTARIO
Weston
Thornhill
Richmond Hill
Victoria Square
Unionville
Markham
Maple
Woodbridge
Pine Grove
Kleinburg
Newton Brook
Aurora
CITY OF TORONTO

### 3. A PROPOSAL TO TAKE TORONTO'S DRINKING WATER FROM THE HEADWATERS OF THE DON AND ROUGE WATERSHEDS, 1887

It's difficult to reconstruct Toronto's watersheds as they were at the time of Euro-Canadian settlement. While contours had to be known for individual engineering projects, systematic mapping of Toronto's relief began only when the federal government started drawing topographic maps in 1904. By that time, many of our streams were long buried, and much of the landscape had been reworked to urban ends. The latter process meant two things: surface contours had been levelled either up or down to produce the neutral grade favoured for property development; and the notion of natural drainage areas had been further eroded by the arrival of sewer systems.

Toronto's natural watersheds gave way to cultural features better described as sewersheds. Thanks to pipes underground and pumping stations on the surface, the City now moves stormwater and sewage across historic watershed divides with relative ease. Watershed maps as recent as the mid-twentieth century have become obsolete, as the Etobicoke Valley in Toronto demonstrates. As late as 1954, its eastern boundary followed a meandering course along the natural height of land. Technology has now pushed the boundary further east, where it runs razor-straight for 5.6 kilometres along Highway 427 and Browns Line. Water that once drained into North and Jackson creeks is now piped into Etobicoke Creek.

Probably the earliest watershed mapping of the Toronto region dates from the late 1880s. For nearly a century after its founding, Toronto had an uneasy relationship with Lake Ontario as its source of drinking water. Purer supplies were seen to be had north of the city, high up on the forested Oak Ridges Moraine – a view that gave rise to an incipient watershed consciousness. In 1886, City Council retained William McAlpine and Kivas Tully to examine alternative water sources. The two engineers tramped across the Don and Rouge watersheds, hatching a scheme to draw water from the Oak Ridge lakes on the moraine and bring it to Toronto by way of aqueducts and conduits. The scheme didn't amount to much – Council came to its senses and accepted the Great Lake on its doorstep as its sole water source – but Torontonians certainly gained an appreciation of the city's watershed context and the (expensive) possibilities of it for water resource extraction.

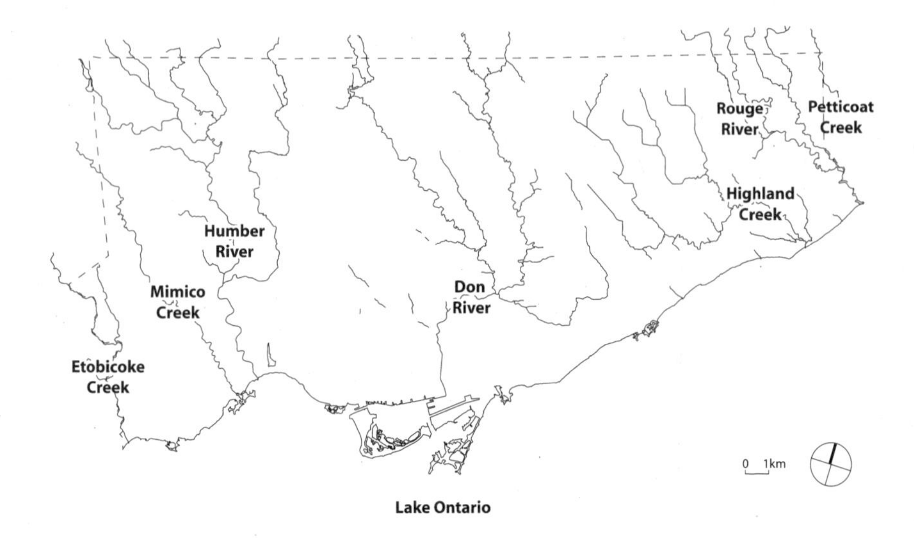


**4. TORONTO'S WATERCOURSES AND BODIES OF WATER, 2008**

The current shape of water within Toronto includes the six waterways that drain to the city's lakeshore (the Etobicoke, the Mimico, the Humber, the Don, the Highland and the Rouge),[6] the ignored or forgotten Petticoat Creek and the Lake Ontario waterfront (including the Inner and Outer harbours, Humber Bay, the lagoons of Toronto Island and the surviving fragment of Ashbridge's Bay). Toronto has 371 kilometres of watercourses and 157 kilometres of shoreline.

Almost invisible at this scale are our inland natural and artificial bodies of water. In terms of surface area, the largest are Grenadier Pond in High Park (19.4 ha), the reservoir in G. Ross Lord Park (14.3 ha), unnamed ponds in Etobicoke's Centennial Park (5.1 ha) and in L'Amoreaux Park (1.74 ha) and Topham Pond in Eglinton Flats Park (1.69 ha). There is also a growing number of much smaller 'stormwater management facilities' (forty-three, according to Toronto Water) and wetland restoration projects that contain standing water at some point in the year.[7]

6 For more comprehensive characterizations of these rivers, see various TRCA watershed strategies and report cards in the 'Protecting Our Water' section at www.trca.on.ca.

7 For a sample of restored ponds and wetlands, call 416-392-0401 for a free Lower Don map produced by the Task Force to Bring Back the Don.

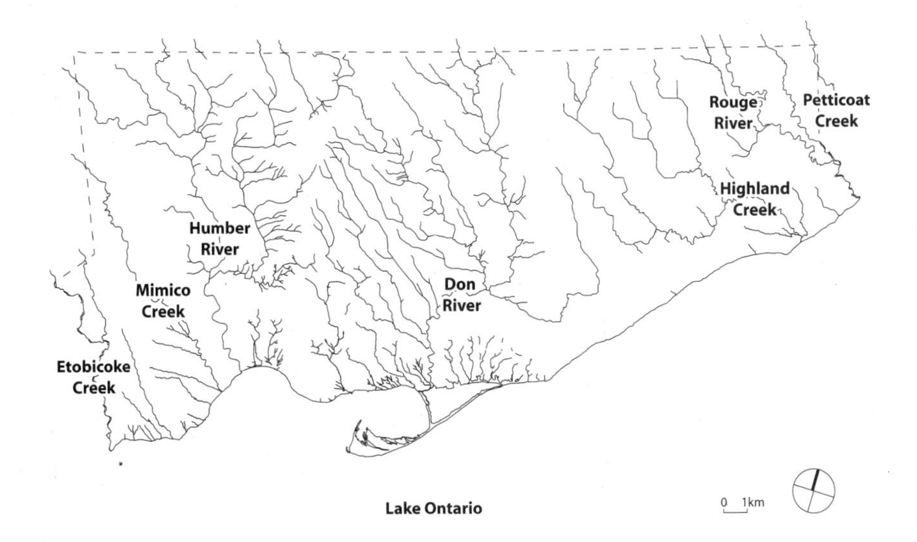


## 5. TORONTO'S ORIGINAL WATERCOURSES AND BODIES OF WATER

Tracing the evolution of Toronto's surface water is easier than charting the changing shape of our watersheds, though the picture gets murkier the further back in time you go. This illustration of 'original' or 'historic' conditions is a compilation of dozens of maps from the late 1780s to the mid-1850s, and draws on the forensic hydrology practiced by the Lost Rivers Project.[8] 'Original' in this sense could be defined as water as we first reliably knew it, which varies greatly as we scan the entire city. No single map tells us what all of Toronto's water features looked like to the British military or Euro-Canadian settlers. Our knowledge of many individual streams is similarly vague and fragmentary; for example, no map exists showing all of Taddle Creek when it flowed on the surface.

At this scale, it's difficult to make out all the manipulations – be they partial channelizations or outright burials – to which our watercourses have been subjected since the early 1790s. Save for the Don River and a tiny portion of Taddle Creek, all of the streams that drained directly into Toronto Bay have now been lost. The Don's tributaries have fared little better, though some fragments of water still flow along the courses of Castle Frank Brook, Yellow Creek, Mud Creek, Cudmore Creek, Walmsley Brook and Burke Brook.

*The past and present shape of water in central Toronto. Original conditions are in grey; current conditions are in black.*

8 See the maps and narratives at www.lostrivers.ca.

**6. THE TORONTO CARRYING PLACE AND OTHER FEATURES ALONG THE HUMBER RIVER, 1933**

Originally published in his 1933 book *Toronto During the French Régime*, Percy J. Robinson's map is a wonderful collage of time and culture, and a bridge between watersheds and watercourses in the Toronto region and beyond.

There are no First Nations maps of earliest Toronto. Our cartographic understanding of Aboriginal life reflects 'discoveries' made by Euro-Canadians beginning in the late seventeenth century. Robinson's rendering of the Toronto Carrying Place from Lake Ontario to the Holland River depicts First Nations' use of the land against subsequent Euro-Canadian trade and settlement. He shows the ancient portage route between the lower and upper Great Lakes and the Aboriginal villages astride it and the Humber River, all in relation to French trading forts and posts, early British survey work and even modern (that is, early 1930s) Canadian communities and roads. The map does not record nineteenth-century milling activities, which, by impounding and diverting water, affected the Humber's shape at the micro-scale. Robinson's work also precedes the extensive flood and erosion control works built after 1954 in response to Hurricane Hazel.

**7. THE DON IMPROVEMENT PROJECT, 1888**

Although Toronto has hatched many big plans for its waterfront and waterways over the past 200 years, most have come to nothing. The Don River is an unfortunate exception. Pitched as a way to relieve the Don's unsanitary state and bring shipping upriver, the 'Don Improvement' actually brought the Canadian Pacific Railway into downtown Toronto and created new land for industry on the flood plain in the late 1880s. Sheet-piling and filling eliminated the Don's meandering course between the Grand Trunk rail corridor and Riverdale Park. Today, the Lower Don Recreational Trail occupies the abortive 'dock reserve' on the west bank of the river, while the Don Valley Parkway hugs the east bank.

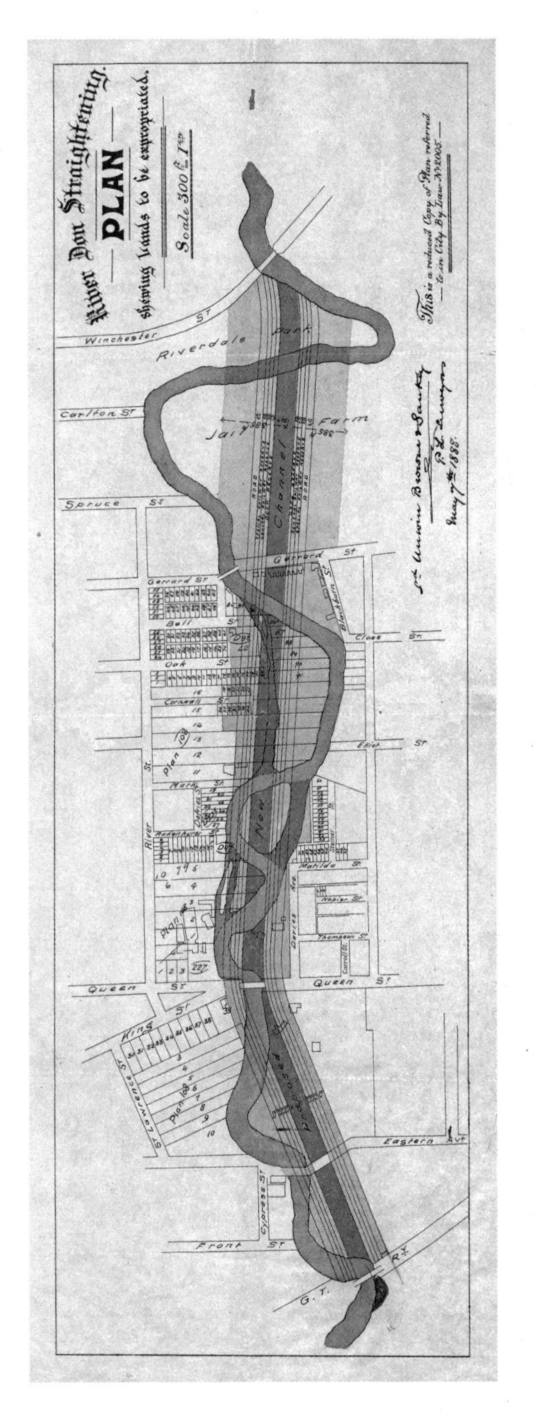

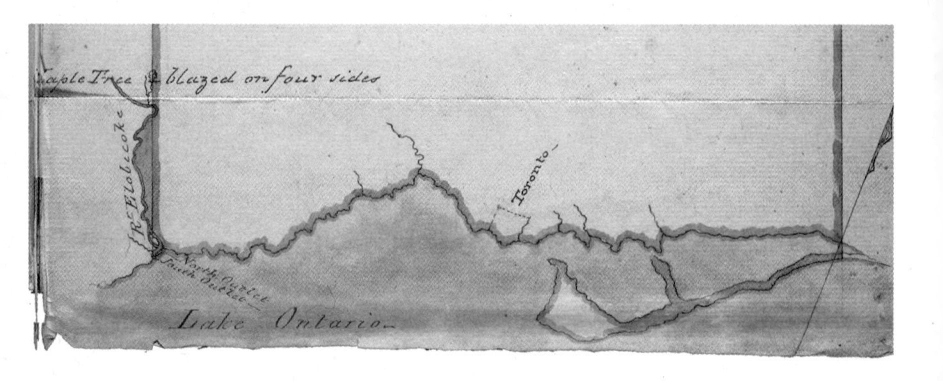


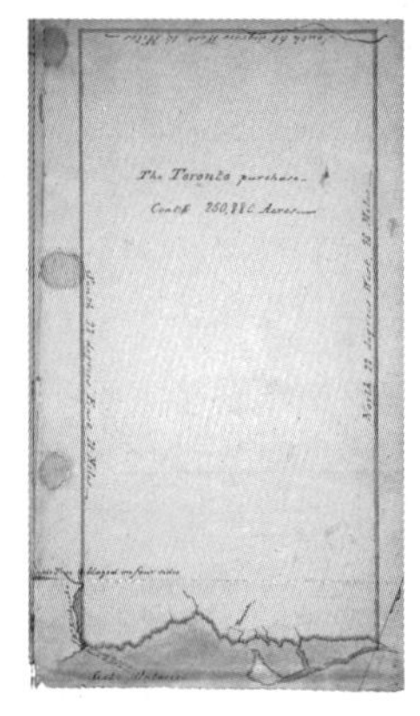

**8. THE TORONTO PURCHASE, 1787**

A final look at Toronto's water harkens back to its first detailed rendering – and a pivotal moment in the relationship between Aboriginals and Europeans and their concepts of property. In 1787, Sir Guy Carleton, 1st Baron Dorchester and governor-in-chief of British North America, cast about for a site that would eventually become Ontario's capital. Dorchester arranged a meeting with three Mississauga chiefs and his deputy surveyor general, John Collins. Out of this meeting came the Toronto Purchase and the Crown's acquisition of 250,880 acres of land.[9] Dorchester and Collins imposed a previously unknown form of order on the landscape: a rectangle perfect on three sides, 22.5 kilometres across the top and 45 kilometres down each of the two long sides. But along the bottom, they could not avoid the ragged reality of the Lake Ontario shoreline.

In a roadless wilderness, Collins began surveying the Toronto Purchase from the southern outlet of Etobicoke Creek. The eastern boundary lay (by accident or design?) at the head of Ashbridge's Bay, near the foot of Woodbine Avenue. In between, Collins mapped the lower reaches of what appear to be Mimico Creek, the Humber River, Garrison Creek, Russell Creek, Taddle Creek and the Don River. The last three streams are shown draining into the north shore of the spit-sheltered Toronto Bay. Collins placed 'Toronto' against the west bank of Garrison Creek, on the site of today's Fort York. When Dorchester ordered up a more detailed survey within the year, the town site found a safer place inside the harbour. This protected water played a key role in the founding of the Town of York in 1793, and in York becoming the capital of Upper Canada three years later. And those decisions led to the inexorable shifting of the shape of Toronto's water.

9 Defects were soon recognized in the Toronto Purchase, though a new deed confirming the 1787 agreement was not executed until 1805. A claim under this treaty by the Mississaugas of the New Credit First Nation is in negotiation with Canada. See the 'Claimsmap – Ontario' section at www.indianclaims.ca.

# Transformations

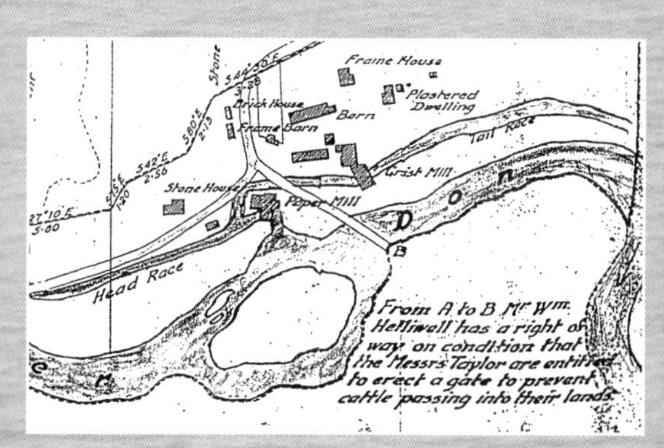

COMMISSION OF CONSERVATION

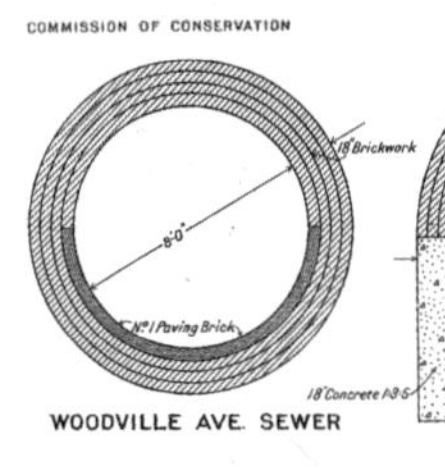


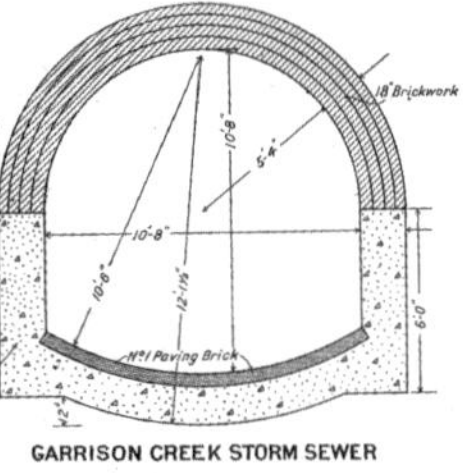


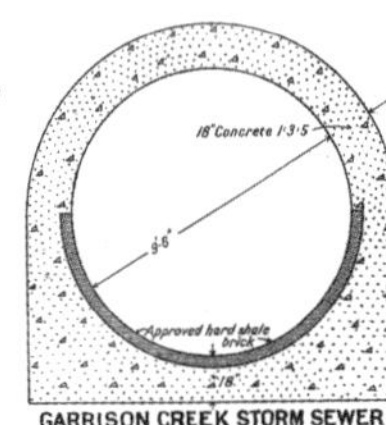


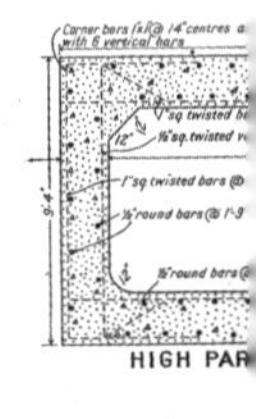

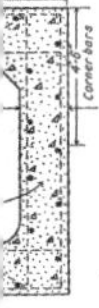

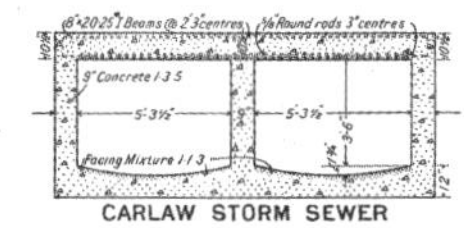


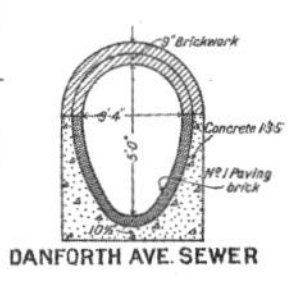


CITY OF TORONTO
TYPICAL SECTIONS OF SEWERS
*Scale 5 feet = 1 inch*

Gary Miedema

# When the rivers really ran: Water-powered industry in Toronto

Twenty-five years ago, I watched my older brother, then maybe twelve or thirteen, build a scale model of a water-powered sawmill. It was an impressive piece of work, complete with a mill dam and pond, a functioning waterwheel and a jigsaw blade. Fill the pond and water turned the wheel that turned a crank that moved the saw. Flick on the electric water pump and the water was returned from the bottom of the waterwheel to the top. The fun could last indefinitely.

My brother's mill testified to the allure of water power, and to the fact that, back then, an Ontario education was incomplete without a lesson on waterwheels, millstones and saw blades, or a class or family trip to a rare surviving old mill. Nearly thirty years later, the model mill still entertains new generations, and old mills remain a quintessential symbol of central and eastern Canadian 'pioneer days,' a staple of Tourism Ontario calendars and fall day trips. Odd, then, that here, in what is now the City of Toronto, we don't often think of our halcyon days of water-powered industry. At the beginning of the twenty-first century, high level bridges, flood-control programs and, in more than a few cases, outright burial have allowed Toronto's rivers and creeks to be conveniently ignored. So, too, perhaps, their rich human stories.

For thousands of years, including at least the first five or six decades of European settlement in Toronto, water routes made this place. Though known by very different names, the Don, Humber and Rouge rivers and Mimico, Etobicoke and Highland creeks had long been important sources of food and means of travel for Aboriginal peoples and, much later, for the early European traders they introduced to the area. With European settlement – and its relentless grid system of roads – the rivers and their valleys received a more mixed evaluation. They now meant extensive detours, steep and difficult hills and expensive bridges all too prone to washouts.

But to European inhabitants of what is now Toronto, rivers and creeks also meant money. Before petroleum and natural gas pipelines, electricity grids or coal-fired steam, water fuelled the nascent industries of Toronto. Water turned wheels that moved

*[1] Mill dams provided a reservoir of water to power mills, protecting them from periods of low water flow in the rivers. Dams also blocked fish migrations, and were in constant need of repair. This 1913 photo captures the decaying Howland and Elliot Dam of Lambton Mills, near Dundas Street, where it crosses the Humber River.*

machinery that produced everything from flour, malt and lumber to textiles and paper. Water, in short, was power.

Water-powered mills and industries had a major impact on the development of the Toronto region. In their day, they functioned as the early-nineteenth-century equivalents of a Home Depot or Loblaws. And, as with today's big-box stores, where mills were built, people would come, sometimes travelling for miles to get there, since mill sites were defined by an often inconveniently located, geographically fixed source of power [1], not by major traffic routes. In the early years of settlement, when roads were mere paths, determined farmers sometimes carried wheat on their shoulders down otherwise impassable muddy trails to get it to the mills. Or they waited for winter, when frozen rivers became veritable speedways for sleighs, which made getting loads to and from the mills an easier task.

The importance of water-powered mills made them the beneficiaries of government largesse, including precious millstones ordered by the Lieutenant-Governor to be given from His Majesty's Stores to encourage the building of the first gristmill on the Don at Todmorden. Almost immediately after the founding of the Town of York in 1793, in fact, the government built its own sawmill on the Humber near the site of today's Old Mill. By 1805, the first sawmills and gristmills appeared on Highland Creek in Scarborough, where the creek was crossed by Kingston Road.

As settlers made their way deeper into the forests and away from the lake, they brought mills with them. Sawmills, often primitive one-storey structures, came first, turning cleared trees into the lumber used to build gristmills when the first crops of grain were ready to be ground. With gristmills came distilleries – a convenient way of turning low-grade grain into another pioneer staple.

Though the lack of surviving old mills might lead you to assume there were never many of them around, that was hardly the case. In the Don River watershed alone, there were twenty-six water-powered mills by 1824. On Highland Creek, some twenty-three sawmills lined the water's edge in 1850. In 1860, perhaps the peak year of water-powered industry on the river, fifty mills churned away. The Rouge, for its part, powered some fifty water-driven mills in 1861.

If water-powered mills were hardly rare, they also defied their often sanitized, nostalgic, modern presentation. Far from quiet places of retreat, water-powered mills were noisy hives of business, and surprisingly large and risky investments. Often run by professional millers hired by a mill owner/investor, mills sometimes changed hands quickly. Specialized equipment for grist or woollen mills was expensive and, at least in the earliest years of settlement, not easy to find. Rivers fluctuated from shallow flows in summer, which rendered mills useless, to dangerous ice jams and raging freshets in spring, which sometimes wiped mills out altogether. Like many other businesses, mills also served temperamental markets. When the British government preferred to import Canadian flour in the 1830s, it became a hot trade commodity, and millers did well. When Britain ended its preferential treatment of Canadian flour in the late 1840s, more than a few mills ended up on the market to satisfy anxious creditors.

The story of Alexander Milne illustrates well the nature of the water-powered mill business. Late in the winter of 1817, Milne, a Scottish-born immigrant to America and a weaver by trade, made the trip from Niagara to York by sleigh. As the story goes, Alexander drove the sleigh while his wife travelled behind his seat, somewhat protected from the elements inside a very large dyeing kettle – a key element of the textile industry. Ten years after his arrival, Milne built his own three-storey mill, Toronto's first woollen mill, on Wilket Creek, a tributary of the West Don River in what is now Edwards Gardens. It was an unfortunate decision: water flow quickly became inadequate to power the carding and pulling machines on the first two floors and the sawmill on the third. Milne persevered, though, moving his whole operation in 1832 to a location on the East Don River at today's Lawrence Avenue. There, a single-storey building handled the textile machinery, and a sawmill was built across the river. Not wanting to be left short of water again, Milne built a dam, which also provided a crossing point over the river.

A temperance advocate and sometime Sunday preacher, Milne wrote in a letter to a friend, 'The only way to live long is make use of our time.'[1] He took that seriously, living to the ripe old age of 99. Second time lucky (or wise), Milne's woollen mill flourished. Where Lawrence Avenue East now rushes straight over the Don, it once meandered south down the slope of the ravine to the valley bottom, passed through a cluster of some sixteen buildings and over a low bridge, then turned north again to climb out of the valley, meeting present-day Lawrence Avenue on the other side. The area that encompassed the meandering road became known as Milne's Mills, or Milford, and included houses for mill workers, a wagon shop and a dry goods store where the family's cloth, blankets, flour and liquor (for temperance did not mean prohibition) were sold.

In 1878, a particularly bad flood destroyed the dam and mills, but Alexander's son William, who had taken over the business, carried on and quickly built a new red- and yellow-brick three-storey mill that remained in operation until 1914 [2]. By then, larger, more efficient, mass-production mills had finally rendered the Milne business uncompetitive and obsolete.

*[2] The picturesque 1878 Milne woollen mill (demolished in 1946) on the east side of the Don River at Lawrence Avenue.*

It's easy to wax nostalgic about mills, particularly those like the Milnes' picture-perfect structure. But water-powered mills came in all shapes and sizes. The Taylor brothers, for example, eventually employed about 100 men at three mills on the Don that produced newsprint for papers across the country [3]. As late as the 1930s, the middle mill near the Leaside Bridge was still powered at least partly by the Don – not by a waterwheel, but by a turbine inside one of the buildings to which water was channelled by a boxed-in flume. It looked and functioned like the factory it was, and not at all like an old mill.

Water-powered mills weren't as 'green' as they might sound to modern ears either. Long before environmental assessments were even dreamed of, mill sites were carved into fragile river-valley ecosystems. With the mills came millraces and, often, dams, which altered the flow of the rivers and creeks to

1 Patricia W. Hart, *Pioneering in North York* (Toronto: General Publishing Company, 1968), p. 75.

*The three Don River paper mills of Thomas Taylor and Brothers, as drawn for advertising purposes in 1877. Some of the Lower Mill site remains intact as the Todmorden Mills Museum. Stacks indicate the use of steam at all three mills, but signs of water power also exist in the large boxed-in flume clearly carrying water to the Middle Mill, and in the diversions of the river through the Lower Mill. Water entered the Lower Mill through the head race, turned an internal turbine or wheel to power the mill, then exited through the tail race, back into the river.*

maximize water speed over the wheel. Early maps of York and Scarborough townships are speckled with mill ponds, one of the largest of which flooded some thirty acres of land. Their dams could also wreak havoc on fish migrations. On Etobicoke Creek, for example, mill dams at its mouth and near Dundas Street were blamed in 1873 for preventing salmon migration. Salmon that made it past dams to their spawning grounds had their eggs suffocated in sawdust or chaff from grain, waste products of the mills that were dumped without thought into the rivers. Waste dyes from woollen mills and waste products of paper mills no doubt also found their way back into the river.

That said, mills built communities. The mills at Todmorden resulted in a 1798 government order to create Mill Road (now Broadview Avenue) to make access to the mill easier from Kingston Road (now Queen Street). At the top of the steep ravine, above the mills, Todmorden developed into a village complete with a well-known hotel and blacksmith shop. Hogg's Hollow (and York Mills) on Yonge owes its existence to mills first established by Samuel Heron and later owned by James Hogg. Lambton Mills on the Humber at Dundas developed around the woollen mills, gristmills and sawmills of William Cooper. Even Malvern had its beginnings as a crossroads town fuelled by the spinoff economies of nearby water-powered mills. In the early nineteenth century, water-powered mills were the seeds from which whole communities sprouted.

At the beginning of a very different century, it's remarkable how little remains of the well over 100 mills that once dotted our rivers and creeks. The fate of water-powered industry was sealed by the depletion of the raw materials it relied on. In 1851, 73 percent of Scarborough was covered by forest. An astonishingly short ten years later, forests covered only 33 percent of the same area, and a census enumerator reported seeing many idle mills, 'the supply of timber having entirely failed.' By 1878, only four sawmills survived from a peak of over twenty on Highland Creek.

Deforestation also significantly altered water tables and the predictable flow of rivers and creeks. Sawmills like Thomas Davidson's, which was located on a small tributary of the Middle Don, employed three people and cut 500,000 feet of lumber in 1851. It was a business doomed to fail. Without forests to shade the ground, water tables fell and, as Alexander Milne discovered at Wilket Creek, rivers shrank to the point that mills could no longer function. At the same time, without trees to help absorb meltwaters and rains, flash floods worsened, carrying dams, and sometimes mills, with them.

The development of other sources of reliable (and geographically unbound) power, specifically coal-fired steam engines and electricity, also worked against mills. The lack of predictable water flow made steam power an early and significant competitor. By 1840, John Van Nostrand had converted his gristmill and sawmill, located under today's 401, west of Yonge, to steam, making them theoretically functional twenty-four hours a day,

[4] *Taken from the east side of the Humber River c. 1895, this photograph shows the Old Mill ruins in their once more remote setting. The bridge, carrying Bloor Street over the Humber, was later crushed by an ice floe.*

365 days a year. By 1850, North York's steam-powered mills were making water power inessential and less convenient. Due to low water levels on the Humber, the Lambton Mills were converted to steam by 1880. Though water-powered mills continued to function through the nineteenth century, often augmented by steam in periods of low water flow, their days were numbered.

Today, it's a challenge to find the remains of more than a few of Toronto's former water-powered mills. Black Creek Pioneer Village boasts Toronto's only functioning example, but it was actually built in 1964 from materials salvaged from an old mill in Prince Edward County. In some cases – take, for example, Joseph Bloor's brewery and millpond, which sat along Castle Frank Brook at present-day Bloor and Sherbourne streets – not only are the mills gone, but the creeks that fuelled them as well. The site of Bloor's picturesque brewery is now run through by Rosedale Valley Road.

Mills and their ponds have experienced mostly unwelcome fates. Floods did their part to wipe out some mill dams, while others were dynamited to prevent drownings after their period of usefulness had expired. Mills have succumbed to fires or have been demolished. The Milnes' 1878 woollen mill stood until 1946, when it was reduced to its three-foot-thick foundations, perhaps for the reuse of its brick. Ironically, conservation efforts have also taken their toll. Ravines and flood plains once stripped of trees have been replanted into forests, while one mill, the one on the Don at the Leaside Bridge once powered by a turbine, was demolished by the Metropolitan Toronto and Region Conservation Authority in 1992, its land immediately integrated back into the surrounding greenbelt.

Redevelopment has also wiped out or greatly altered many mills, including the well-known Old Mill on the Humber, just north of the Bloor subway crossing [[4]]. One of the best known remnants of Toronto's early mills, its vacant, five-storey stone structure was a tangible link to the past of Toronto's oldest industrial site, which boasted a mill as far back as 1793. The rare ruins were the shell of the fourth and last flour mill there, built for William Gamble in 1848. Gamble's mill was used only until 1858. It suffered a fire in 1881 and was thereafter abandoned, but the mill's massive walls would stand for over 150 years, a haunting reminder of a day when water drove the wheels that put bread on the table and when river valleys were sites of industry as well as recreation. The ruin even survived the legendary floods caused by Hurricane Hazel in 1954. In 2000, however, an agreement was reached that allowed the dismantling of the ruins and the construction on its site of a new boutique hotel loosely modelled after the mill and partly clad with its stones.

Yet if you look carefully, intact hints of the old mills can still be found. Though much has changed where Lawrence Avenue crosses the East Don River, a road still winds south down the east bank into the ravine and a City park called Milne Hollow, where one Milne home still remains from Milford's glory days. Remains of an old millrace from Wadsworth's Mill can apparently be seen running across the second fairway of the Weston Golf and Country Club. At Lambton Mills, you can still follow the old line of Dundas Street down to the water, as it passes a rare remaining building from the village's milling heyday, the late 1840s Lambton House. Walk south from there down the Humber and the foundations of Thomas Fisher's 1830s mill can still be seen behind shrubs and trees on the west side of the valley.

In fact, if you poke around a bit, who knows what you'll find. In 1948, a bulldozer digging into a hillside off Pottery Road unearthed five millstones, perhaps those of the first mill on the Don, established there some 150 years before. If you don't have a bulldozer, maybe find someone who has a membership at the private Donalda Golf and Country Club, located southeast of the corner where the aptly named Don Mills and York Mills roads intersect. Play some golf and, on the sixth tee, wander over to have a look at the nearby barn buildings. One of them is the altered remains of William Gray's gristmill, built in 1828 and converted into a barn long before the days of the country club.

Those scattered remnants of mills aren't much for a landmass as large and as well-watered as Toronto. But given our long propensity for rushing into the future without too many glances back, we're lucky to have these fragmentary reminders of our long and often turbulent relationship with water.

SELECTED SOURCES


City of Toronto and Toronto and Region Conservation Authority, *State of the Watershed Report: Highland Creek Watershed* (Toronto, August 1999).

Etobicoke and Mimico Creek Watersheds Task Force, *Greening Our Watersheds: Revitalization Strategies for Etobicoke and Mimico Creeks* (Toronto, May 2002).

Patricia W. Hart, *Pioneering in North York: A History of the Borough* (Toronto: General Publishing, 1968).

Ontario Department of Planning and Development. *R.D.H.P. Valley Conservation Report, 1956* (Toronto, 1956).

Ontario Department of Planning and Development, *The Humber Valley Report* (Toronto, 1948).

Ontario Department of Planning and Development, *The Etobicoke Valley Report* (Toronto, 1947).

Ontario Department of Planning and Development, *Don Valley Conservation Report* (Toronto, 1950).

Charles Sauriol, *Pioneers of the Don* (East York, Ont., 1995).

Richard Anderson

# The dustbins of history: Waste disposal in Toronto's ravines and valleys

Like most cities, Toronto is built on its waste materials. It produced very little rubbish in its early days because it was small and burned wood. But from the 1850s, when it became a large, coal-burning city, Toronto began to generate waste on a much vaster scale. Creeks and water bodies came in handy as dumping grounds, and dozens of them vanished. Houses, apartment buildings, parks and schools have been constructed on their filled remains.

We have mostly ignored these filled ravines and lost watercourses. These dustbins of history remain outside the public consciousness, out of sight and largely out of mind. But every so often we get a little reminder, a sign that lost creeks are still there, and their legacies still lurk in the landscape. One of these moments came in the summer of 2001 when two stories gained the attention of the city's media.

In August, the residents of Swansea learned that elevated levels of lead had been found in the soil of their public-school playground, once part of the Rennie Creek ravine. About a month later an environmental group, Lake Ontario Waterkeeper, found toxic water seeping into the Humber River from King's Mill Park. Testing revealed large excesses of ammonia, polychlorinated biphenyls (PCB) and polycyclic aromatic hydrocarbons (PAH). The ammonia levels alone were enough to kill fish. None of the samples met Ontario's clean-water guidelines. A little research showed that old landfills were responsible for the contamination in both cases [1].

It is embarrassing to contaminate Toronto's only federally designated Canadian Heritage River. It is equally embarrassing if you appear to expose schoolchildren to lead in a playground. Sympathetic city officials expressed much concern and acknowledged the presence of old dumps in both cases, although they were hazy on the historical details. For one reason or another, the hazards of these old landfills had simply escaped official attention. Apart from some methane venting in King's Mill Park, neither site was being monitored. The City didn't know about the Humber seepage until Waterkeeper told it.

Both these cases are now into their remediation phase. Fixing these problems is a slow and expensive business, and

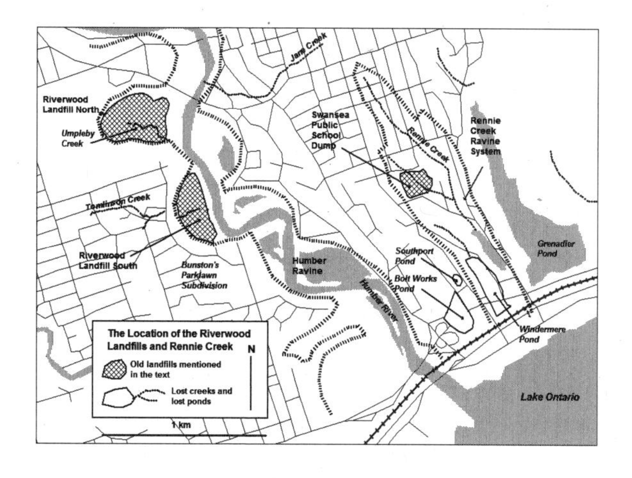


**1** *The location of the Riverwood landfills in King's Mill Park and dumps in the Rennie Creek ravine.*

resources are always in short supply. Cutbacks in provincial spending in the 1990s and the downloading of costs onto the City government have not made this easier. Whatever the remedies available, and whatever the frustrations of applying them, these two incidents share common themes. They show us some of the texture of urban environmental history, and the unfortunate consequences of losing our grasp of it.

THE RENNIE CREEK RAVINE Swansea is a fine community by all accounts, with some of the city's loveliest residential areas. These are, unfortunately, intermingled with some of Toronto's most intensely filled ravines. Swansea Public School was built on the western rim of one of them, the lost ravine of Rennie Creek [2]. A deeply incised ravine, it contained a small river system that drained (or drains) into Lake Ontario. Railway construction in the 1850s disrupted its outlet and created a series of ponds, though the ravine's steep slopes discouraged urban development until the twentieth century.

As the city grew and Swansea became a suburb, municipal authorities began to use the Rennie Creek ravine as a conveniently located garbage dump. It was an empty void near a district whose residents wanted York Township to collect their garbage. In April 1914, the Beresford Dump began to take ashes, rubbish and garbage from the surrounding areas. Dumping expanded and, by 1920, the back of Swansea Public School accommodated an active and disgusting open dump. The 1920s brought flies, odours, rats, blowing papers and, above all, smoke from the frequent dump fires. The dump smouldered

**2** *Swansea Public School yard, with Rennie Park in the background, 2008.*

for weeks at a time, its smoke and smell often preventing residents from opening their windows in hot weather. Crickets from the dump invaded nearby homes in summer, devouring wallpaper and clothes. Stung by public criticism, the school board moved to develop the site as a playground, a process that involved levelling, grading and some judicious dumping of additional materials. In 1931, acting in his capacity as school-board secretary, David Crombie Sr. issued the debentures necessary to complete the filling and build the playground. It took years to finish and wasn't ready until the 1940s. Even then, glass and metal objects kept bobbing to the surface of the fill. Those who attended the school in the 1950s can remember the spring student routine of a fingertip search of the playground to find the winter's sharp heavings.

As operations at the Swansea school dump were scaled back, dumping shifted to other parts of the Rennie Creek ravine. The upper portion of Rennie Park was worked as a dump in the 1930s by local men on relief, which is how it got the nickname Pogey Park. A southern branch of the ravine was developed into the massive Windermere Dump in 1929. Operated by the City of Toronto and the Village of Swansea, it accepted 500,000 tonnes of the city's domestic and industrial wastes until 1953. After the Windermere Dump closed, a remaining section of the Rennie Creek ravine below Pogey Park was used as a Village dump. This sanitary landfill operated from 1953 to 1966 and used a bulldozer to crush and bury wastes. In a burst of municipal pride, the Village allowed its bulldozer to feature in several 1950s heavy-equipment commercials as it clanked about. The completed landfill became part of Rennie Park.

KING'S MILL PARK The origin of the King's Mill Park landfills is similarly intricate. The Riverwood landfills, as they were known, were located on the flood plain of the Humber River, on the west bank, below Bloor Street. The landfills occupy two marshy oxbows, old segments of the river's meandering channel that were abandoned by its changing course. In the 1950s this reach of the Humber was an active flood plain, with oxbows, marshes and braided channels. It flooded frequently, especially after heavy rain or in spring breakup, when ice dams forced the river over its banks. Ice jams in 1922 brought flooding as far up as the Old Mill, uprooting logs, shacks and large trees. Another jam in January 1930 destroyed a squatter's shack on Mudbank Island, just below the Old Mill Bridge. And Hurricane Hazel was simply the most disastrous of several Humber floods.

Although erosion had silted the river to half its natural depth and sewage had reduced its water quality, the lower Humber still behaved like a natural river in the 1950s. Its waters took several days to build to flood peak after heavy rains and it sustained fish catches, even if it sometimes smelled like a sewer. It was a favourite spot for catching sucker fish in the spring, despite water-quality declines from the mid-1920s, when Toronto's outer suburbs began to dump partly treated effluent upstream. For the next forty years, that sewage prevented the Humber from freezing in winter. But the river was far from dead, and in 1959, when Toronto's officials prepared to dump at Riverwood, they found active wetlands on the site. There were marsh plants, mosquitoes, dragonflies and other insects. Small watercourses drained the soggy ground. Despite all that had happened, the Riverwood oxbows still had plenty of ecological vitality.

The lower Humber had long been a recreational spot for the surrounding areas. From the 1860s until well into the twentieth century, it attracted summer excursionists from Toronto. On the shady banks of the Humber, people could escape the city's heat and smoke. They could picnic, hike, fish, swim, wash the car and canoodle. The nude bathing scandalized local moralists but, despite several drowning incidents and overzealous policing, bathing continued there until at least the 1940s. A hot weekend in the 1930s would bring 1,500 to 2,000 bathers to the lower Humber to frolic in the diluted sewage. Naked teenagers romped in the shallows and giggled in the trees, quite beyond adult control.

But the lower Humber was being gradually enveloped by the sprawling metropolis. Relatively unspoiled until the 1940s, it attracted upscale residential development in Swansea, on the east bank, and in the Home Smith estates on both sides of the river. Providing for these homes and accommodating suburban needs required environmental alteration. The minor creeks and ravines on the Swansea side were obliterated by landfill in the 1920s and 1930s, a process that had extended to the Etobicoke side by the late 1940s. Clint Bunston, one of the local property kingpins, was an ambitious user of landfill for real estate development. The 'Deposits of Fill' records in the City of Toronto Archives show him ordering thousands of truckloads of the city's waste to fill up ravine lots in Swansea and High Park in the 1940s. After 1945, he turned his attention to the west bank of the Humber. To achieve his development aspirations and those of his clients, Bunston arranged for city ashes and rubbish to be dumped in the area south of Bloor. A four-hectare, ten-metre-deep ravine was filled to accommodate construction in 1948 and 1949. The ravines of two Humber tributaries were filled south of Sunnylea Avenue in the early 1950s; one of them received 5,000 truckloads of waste. In 1952, Bunston launched an even grander scheme: a thirty-nine-hectare, $7 million subdivision of 1,480 buff-coloured low-rise apartment units east of Park Lawn Road and north of the Queensway. More fill was dumped, more ravines obliterated and more streams buried in sewers.

By the mid-1950s the two Riverwood oxbows were inviting voids in the suburban landscape. Thanks to Bunston's dumps, Toronto's Department of Street Cleaning had come to know the area intimately. Bunston had even offered to sell nine hectares of the northernmost oxbow to the City as a dump in June 1952, but the deal fell through. The oxbow was in Etobicoke Township, a rival municipality then beyond the City of Toronto's limits. But the post-war economic boom and changing waste generation made the city's shortages of dump space critical. By September 1954, Toronto's major west-end dump, Windermere, had closed and the major east-end one, Greenwood, was about to follow.

Salvation, for the City of Toronto, came in the form of Hurricane Hazel. Hazel ravaged the area in October 1954, transforming the politics of waste disposal. The storm hit the Humber especially hard, washing out unwisely sited subdivisions and sewage plants. Flood-proofing the Toronto region became a megaproject involving four levels of government and the expropriation of 500 homes in the danger zones. Lowball expropriation proceedings eliminated obstructive landowners, while the gravity of the situation overcame petty municipal rivalries. The flood-liable areas would be converted into parkland, the river channels were to be re-engineered and the flood plains filled to raise low-lying ground above danger levels. These fills would also accommodate the huge volumes of waste being generated by the booming Metro Toronto economy.

Another megaproject, laying trunk sewers down the valley to a new Humber Wastewater Treatment Plant near the lakeshore, would remove suburban sewage from the river.

So it was that the Metro Toronto government, then lacking powers of waste disposal, embarked in 1955 on a regional landfilling scheme, selling it as a flood-proofing and park development project. Known as Operation Overload, it created landfills deep in the city's major ravines and directly adjoining its major watercourses. By 1964, Metro was handling half the region's solid wastes and most of its industrial refuse.

Metro designated the two Riverwood landfills as Land Improvement Scheme #5. Riverwood North was 9.5 hectares, while Riverwood South was smaller, at 5.7 hectares [3 4]. They were intended to accommodate Toronto's most pressing contemporary garbage problem: industrial waste. In the rhetoric of the Metro Works Department, these dumps weren't for ordinary garbage, but for 'selected' trade waste. Like the other Overload schemes, these were to be sanitary landfills. Waste dumped each day would be crushed and covered with 'clean' fill, bulldozed into place. The usual nuisances of open dumps – the rats, the odours, the flies, the smell, the blowing papers and the smoke – would be eliminated. This progressive state of affairs also allowed the dump to be operated less than 100 metres from new housing projects.

3 LEFT *Riverwood South, looking north from the Bunston subdivision across the Humber oxbow, with Tomlinson Creek winding somewhere through the meadow. This photo was taken November 9, 1959, immediately before landfilling started. The access road to the Humber Yacht Club is visible against the northern cliff of the oxbow.*

4 RIGHT *Riverwood South, looking south, before Metro Toronto began landfilling. This photo was taken November 9, 1959. Some illegal waste dumping had already occurred, and an apron of fill material had already been laid to provide site access. Several of the buff-brick mid-rise apartments of the Bunston subdivision are visible behind.*

Access roads to both oxbows were finished by April 1959. To keep the Humber out of the fill, Metro built perimeter berms and arranged with Etobicoke to divert surface drainage. The remains of Tomlinson Creek, already abused by Bunston's dumping, were channelled into a storm sewer across Riverwood South, while another drain was constructed to divert Umpleby Creek across Riverwood North. Metro brought in a dragline and bulldozers and contracted with Dumps Industrial Ltd. to salvage scrap metal and paper.

In late November 1959, both sites began to receive large quantities of industrial waste. In 1961, a trade publication, *Canadian Municipal News*, feted the sites. The reading public was told that 600 trucks a day were dumping at the site (it was closer to 600 a week). Metro was weighing the trucks, but it was still customary to count truckloads rather than to quote tonnage figures for waste quantities. Metro Works Commissioner Ross L. Clark, standing like a trousered colossus, posed proudly for the photographers in his white hard hat. The publication captured trucks of varying size, shape and state of disrepair using the landfill. A few were the shiny new private packer trucks of Metro's emerging private waste haulers.

Not all of the publicity was good. While the trades presented Metro in a glowing light, nearby residents still complained of odours, flies, rats and blowing paper. The local newspapers picked this up. Metro officials brushed the complaints off in functionalist style: the inconvenience, said Clark, would be temporary, but well worth enduring if Metro could turn a useless swamp into a park.

Originally scheduled to operate from 1959 to 1961, both of the Riverwood landfills accepted materials until 1963 and did not complete their close-out cover until 1967. Their fifteen hectares eventually accommodated 375,000 tonnes of industrial waste plus cover material. Although benign-looking and park-like today, the dumps were just a few metres from the river and lacked structures to contain seepage. For more than forty years they bled leachate into the Humber River unnoticed.

In the *longue durée* of Toronto's waste disposal, the landfills at Swansea Public School and King's Mill Park are just two symptomatic episodes. Instances of filling similar to these have been repeated across Greater Toronto. We have, in the past, routinely buried wastes in ravines, water bodies and creeks. Such spaces were often disregarded as waste ground in the past, and filling them seemed an improvement. Hundreds of Toronto dump sites now lie scattered across the historic record [5]. The present understanding is that most of the old coal-ash dumps are not much of a toxic threat, but some contain unpleasant materials. Most old dumps lack the means to control or monitor ongoing pollution, seeping their contents unheeded into surface and

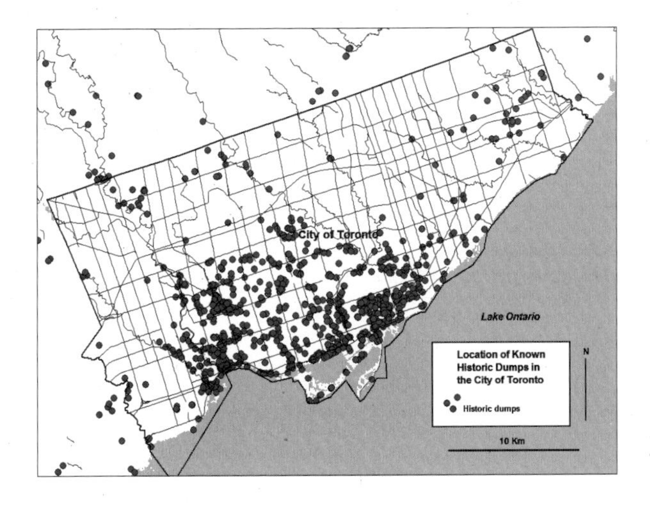


5 *Location of known historic dumps in the City of Toronto.*

groundwaters. They vent their methane slowly and sometimes explosively. We have built on them and we continue to do so.

Public officials show some awareness of the problem of old dumps and filled ravines, but are often vague on the historical facts. Despite their history of municipal use, many landfills have passed out of institutional memory. The Ontario inventory of waste-disposal sites, for example, lists eighty old dumps in the City of Toronto. The two King's Mill Park sites are in that inventory (if we overlook their incorrect map references), but the dumps of the Rennie Creek ravine, and those undertaken by the indispensable Mr. Bunston, are missing. Even the massive Windermere Dump is not in the provincial inventory, and it was bigger than the two in King's Mill Park. If you check with the City, you find that it acknowledges some of the sites that the Province does not, but the information can be very patchy. There is also a great deal missing. To the eighty dumps recognized by the Province within Toronto, I would add another 600 for which I have documentary evidence, and that probably isn't the whole picture. Most of the missing sites are small, and many are probably innocuous, but not all. To manage the hazards, we must first identify the sites.

As the residents of Swansea and the fish of the Humber have discovered, amnesia carries risks. We need to figure out what we have dumped into Toronto's ravines, and where. We must improve our grasp of our city's environmental history if we want to live sustainably within it.

Chris Bilton

# Storm warning: Hurricane Hazel and the evolution of flood control in Toronto

The island of Grenada is nearly 4,000 kilometres from Toronto, its tropical isolation another world away, especially when compared to the chilly, rain-soaked fall weather Toronto was enduring at the beginning of October 1954. When the eighth tropical storm of the year whipped itself up just off the Grenadian coast on October 5, the disturbance was still about a week away from a destination as unlikely as Greater Toronto. News of the storm barely registered in the city, where any major disturbances usually succumb quickly to the vast stretch of Canada's eastern land mass. But Hurricane Hazel would soon make a lasting impression on Toronto.

After devastating Haiti with 185 km/h winds, massive tidal waves and horizontal rain, the storm surged toward the eastern U.S. seaboard. Maintaining equally powerful winds and rising tides a full ten feet above normal as it made land in the Carolinas, Hazel then swept north through Washington, D.C., and into New York State, where it was taken off hurricane class as it approached the Canadian border. Hazel would, however, regain its momentum as Lake Ontario provided the nourishing moisture needed to re-energize its destructive fury.

At 11:10 p.m. on October 15, a reinvigorated Hazel tore through Greater Toronto, savaging the region in a way no one could ever have imagined. So unprepared was the city that most downtown residents in the city's core neighbourhoods slept peacefully through what they assumed was just another bad thunderstorm, while only a few kilometres away entire neighbourhoods were being devastated. The 'wall of rain,' as some described it in Betty Kennedy's book, *Hurricane Hazel*, dumped approximately 40 billion gallons of rain into Toronto's already-swollen rivers, which then 'chewed at the riverbeds and bridges,' not only flooding communities but, in one case, washing away the whole of Raymore Drive and killing thirty-eight residents [1]. A veritable lake consumed the low-lying neighbourhood in Eglinton Flats, and at least twenty of Toronto's bridges were destroyed or severely damaged. Flooding from the nearly forty-eight hours of constant rain and one impossible hurricane accounted for eighty-one deaths and over $20 million in damages across the Toronto region.

▪1 *Rescue efforts during Hurricane Hazel, October 15–16, 1954.*

▪2 *The devastated Long Branch flats area, near the Etobicoke Creek mouth, 1954.*

No natural disaster commands such instant recognition in Toronto as Hazel. Even a half-century after the massive storm slammed down in south-central Ontario, the name is synonymous with unprecedented loss of life and widespread destruction. The high-water mark still visible on the Bloor Street Bridge over the Humber River is a staggering twenty-five feet up. In Toronto, the areas hardest hit were those along the city's major waterways: the Humber River through Weston Road, around the Old Mill area and along the length of Black Creek; Etobicoke Creek at Long Branch flats [2]; the Don River at Hogg's Hollow and through parts of Don Mills Road. Not even the Great Fire of 1904 had so thoroughly crippled the city.

Toronto had long experienced flooding during spring thaws and occasional storms that set the waters surging. Even after earlier storms of great severity, including one in 1878 that was also categorized as a natural disaster, little was done in response to these surges aside from improving methods of rebuilding. In 1850, when the Humber rose six metres and washed away or badly damaged every building in Weston – then situated on the lowlands on the west side of the river – the community met, abandoned the flood plain and resettled high up on the east bank of the Humber, where it's been ever since.

But as Toronto's growing population encroached on the unpredictable flood plains, the relationship between the city's waterways and its citizens needed re-evaluation. Prior to Hazel, the four Toronto-region conservation authorities that had been established since 1946 had taken stabs at flood control: the Etobicoke and Mimico Conservation Authority constructed channels in Long Branch and Brampton; the Humber Valley Conservation Authority planned a dam and channel improvements at Black Creek; the Don Valley Conservation Authority undertook a public education campaign; and the Rouge-Duffin-Highland-Petticoat Conservation Authority acquired floodplain property in Highland Creek. After Hazel's wake-up call, however, planners began to think about flooding in terms of a range of preventative measures on a massive scale. Thus began the evolution of flood control in Toronto.

Achieving a broad regional perspective to address flooding meant the amalgamation, in 1957, of the four existing conservation authorities to form the Metropolitan Toronto and Region Conservation Authority.[1] Championing this transition were Frederick Gardiner, the first Metropolitan Toronto chairman; A. H. Richardson, the provincially appointed chairman of the TRCA; TRCA field officer K. G. Higgs; Etobicoke and Mimico Authority treasurer Frederick Lunn; and watershed-engineering guru G. Ross Lord. The organization also had the financial backing of the Ontario government and Metro Toronto.

The 1957 restructuring led to flood-protection works that mirrored heroic post-war infrastructure projects like the Yonge Street subway, the Gardiner Expressway and the social housing at Regent Park. The new regional body also provided residents and politicians with the opportunity to see Toronto's waterways as part of a system whose every aspect would contribute to a better understanding of how to prevent the kind of tragedy Hazel had wrought. But such an undertaking proved incredibly challenging, even before a plan of action was envisioned.

1 The MTRCA became the TRCA after Toronto's amalgamation in 1998. For simplicity, 'TRCA' will be used here regardless of the date being referred to.

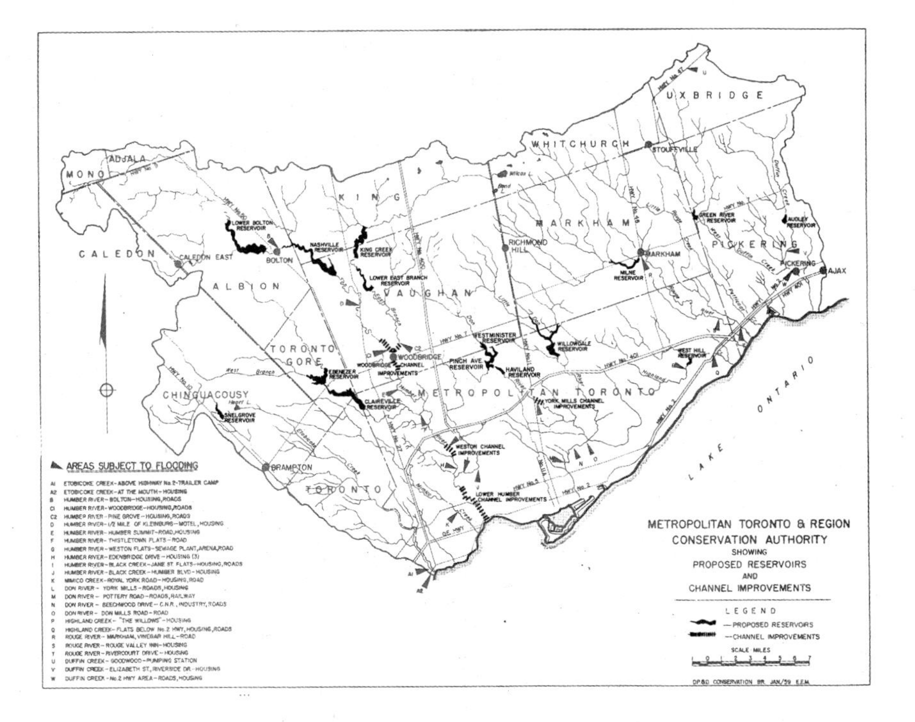


Not only were the efforts of the TRCA the first grand-scale attempt at preventative flood-control measures, the concept of flood control itself was so new that even the most basic practices for tracking and measuring fluctuating water levels had to be developed from scratch. Fortunately, as detailed in former TRCA chief administrative officer Bill McLean's book, *Paths to the Living City*, a generous supply of forward-thinking engineers, combined with a profound sense of mission, provided the authority with workable resources within the first year of its existence.

*[3] The Plan for Flood Control and Water Conservation, 1959. Projects along the Black Creek are not shown on this plan.*

The result was 1959's *Plan for Flood Control and Water Conservation* [3]. It called for the construction of multipurpose dams and reservoirs, river-channel improvements, the acquisition of flood-plain lands and the creation of a flood warning system over a ten-year period – heroic feats of engineering intended to check the destructive forces of nature. The plan also set down an underlying theory of flood control:

> For the most part, the various measures proposed for each watercourse are inseparable from one another. Channel improvements are complementary to the reservoir

> systems.... The acquisition of flood plains reduces the size of required reservoirs, permits the utilization of less extensive protective works, and also forms an integral part of ancillary conservation measures [such as recreation].

In short, for this plan to work, all of its components had to work together.

If the scope of the plan was ambitious, the proposed major projects were nothing short of iconic. Engineering dams right in the middle of the city were not only an effective flood-control measure, it was also one that everybody could see. Since the goal of the authority's initiatives was for nothing to happen – no floods, no destruction, no death – having something to show for the money spent on the projects was the only tangible way to demonstrate progress to the public.

Initially, dams were the favoured method for controlling the rising waters of Toronto's rivers and creeks. Flooding along Black Creek (a major urban tributary of the Humber) had long been a concern, and planning for control works was well underway before the TRCA was established. As a result, Black Creek was named as the highest priority in the TRCA's plan. The small retardation dam in Downsview Dells Park, completed in 1960, became the TRCA's first dam-building project.

Heavy engineering entailed a vast amount of construction spread out across the Toronto region. Sixteen dams and reservoirs were proposed, including seven within the city. The West Don River would host a trio of dams, dubbed the Haviland, Westminster and Finch dams, extending northeast from Dufferin

■4 *The G. Ross Lord Dam and Reservoir in G. Ross Lord Park, 2008.*

and Finch. The East Don would have the Willowdale Dam. The West Hill Reservoir was slated for Highland Creek. The Humber would have the Claireville and Ebeneezer dams on the city's northwest border, plus the Black Creek retardation structure. Major channelization works – a mix of concrete lining, weirs, dikes and protective rip-rap – were also planned for the lower Humber, Weston and York Mills, in addition to Black Creek.

The TRCA's other major project involved a softer approach. Flood-plain lands were acquired so the risks of flood damage could be transferred from private owners to the authority, thus hopefully also eliminating recurring death and destruction, and allowing Metro Toronto or the authority to develop the lands as parks and conservation areas. As the plan asked, 'In a highly developed urban area such as Metropolitan Toronto and Region, where the main water sources vitally affect the health and social well-being of the entire populace, should the lands along those major streams be considered as public domain?'

In 1959, some 7,200 acres were proposed for acquisition; this target soon grew to 22,000 acres. In Toronto, the conversion of flood-plain lands into parks was confirmed under a 1961 agreement between the TRCA and Metro Toronto. As a result, the TRCA is now the underlying landowner of 51 percent of the City's parks system.

The TRCA's acquisition ambitions and achievements were only slightly less impressive than the engineering feats it managed to accomplish. Though six large multi-purpose dams were initially slated for construction in Toronto, by 1962 some of the smaller

[5] *The channelized West Don River at York Mills, 1995.*

ones had been deemed ineffective and unfeasible in their locations. In the end, only two were actually completed in Toronto. The first was Claireville in 1964.[2] The second was the Finch Dam at Dufferin and Finch, renamed the G. Ross Lord Dam and Reservoir after the head of flood-control development, which was increased in size to support the capacity of feeder streams and to stem the chronic flooding of the West Don. Completed in 1973, the G. Ross Lord reservoir serves as Toronto's second-largest inland lake [4]. However, the planned recreational functions attached to that dam were abandoned due to pollution, fluctuating water levels and the potential instability of the embankments.

Engineering feats were also evident in the channelization work completed to improve the passage of floodwaters and prevent riverbank erosion common during flooding. The largest project was at Black Creek, which flows in concrete through a 7.2-kilometre stretch between Jane and Wilson and the Humber. Similar treatment of a 1.6-kilometre section of the creek at York Mills addressed the drainage problems posed by an eighty-eight-square-kilometre area [5].

While the dam construction, the land acquisition and the channel engineering were going on, the TRCA had to deal with a different kind of flow – cash flow. When the plan for flood control was first developed, it carried a price tag of $40 million spread out over a decade. But in the wake of tragedy brought on by Hazel, the TRCA was able in 1961 to convince the provincial and federal governments to split 75 percent of the cost, while the municipalities picked up the remaining 25 percent. It was a bold gesture, and possibly presumptuous, but it appeared to be the only way any of the major recommendations would be implemented.

Even with financial support from all governments, the rising cost of construction and real estate (even for land that would be otherwise undevelopable) meant that things were already looking to go way over budget by the time the TRCA issued a five-year progress report in 1966. For a time, funding was extended. But that's not to say things were looking up.

In *Paths to the Living City*, Bill McLean notes how, with the cost of the project continuing to rise and completion dates stretching farther and farther away, the TRCA was subjected to shifting trends in provincial and federal policy as new political parties took control. The TRCA also had its own growing pains to deal with, namely the direction it should take with certain

2 While the dam lies within Toronto, virtually all of Claireville Reservoir is in Peel Region.

aspects of the project, the findings of new research and reflection on its accomplishments to date.

All these factors came to a head in the late 1970s, when the harsh reality of a recession only served to further underline the fact that the flood-control plans were topping out at over double the projected cost. A new provincial agency, the Ministry of Natural Resources, was created with the responsibility of overseeing Ontario's conservation authorities. In an effort to reduce costs and streamline the authorities, the MNR sought a new direction for the authority that steered it away from flood control, suggesting it should focus on specific programmatic aspects of watershed conservation like shoreline regeneration, each aspect subject to its own financial and regulatory constraints, rather than to view things through the flood-control lens, where all aspects were connected.

By 1976, the data and analysis used for the 1959 plan were outdated, and a complete review of the plan coincided with the shifting power structure. At the same time, growing criticism that the TRCA lacked interest in community involvement forced it to consider outside opinion. The authority's heavy-engineering focus came under criticism for its high cost; resource management and a policy of 'non-use and non-intervention' was now preferable in the rapidly developing urban areas. It was here that the TRCA reluctantly transitioned into a role closer to what we see today. With less money available and the MNR exerting policy influence, the TRCA evolved in order to maintain its original objective of flood prevention. Watershed management became the focus.

In the beginning, the TRCA had assumed jurisdiction over what some considered an 'ecological slum,' according to *Greening Our Watersheds*, its 2002 report. Many of the rivers contained alarmingly high levels of pollutants. Toronto, like every other North American city, had been treating its waterways as a natural system for removing waste and sewage for over 100 years, and both aquatic and terrestrial habitats were degraded or faced the prospect of being lost altogether. So, in addition to its flood-control efforts, the TRCA's emerging concerns for ecology and watershed health made the organization look like a de facto environmental advocacy group.

Conservation had always been a central element in the TRCA's approach to flood control. By acquiring flood plains and preventing development in and around those areas, they had already taken the first steps toward a more ecological management method. From the late 1970s onward, a different approach to the watersheds, and in turn flood control itself, became necessary due to the constraints facing the TRCA. (Only 40 percent of the 1959 plan had been completed by 1979, and the prospect of more heavy engineering looked unlikely.) When the authority finally adopted its Watershed Plan in 1980, it had officially re-evaluated its understanding of the 'ecology of valley resources.' In the plan, it committed to addressing all aspects of the natural environment within the flood plains and along the ravine sides while still ensuring protection from floods.

One of the most influential drivers behind this new approach to conservation came from the very thing that flood-control works were protecting: the people who lived near Toronto's waterways. Groups like Save the Rouge Valley System, the Task Force to Bring Back the Don and Action to Restore a Clean Humber used volunteer efforts and community activism to highlight environmental concerns in their watersheds. At a time when financial and power structures within the TRCA were less than stable, these groups were (and still are) able to bolster support and bring local focus to big issues. Water quality and aquatic life had become at least as important as water volume.

Appropriately, one of the TRCA's current major projects is the Don mouth regeneration, which involves both ecological recovery and flood control. Habitat creation is prominent in this project and is symbolic of the TRCA's new direction, but flood control remains imperative. Without a new 'flood protection landform' on the Don's west bank, there's no possibility of developing the West Don Lands for residential or commercial use, or permanently protecting other downtown areas from flooding. In a nod to the TRCA's past, heavy engineering is required to realize these objectives. But as an evolution of engineering design, the entire Don River Park will serve as a flood-control structure, using the most elegant berm Toronto's ever seen to dispense with the concrete and rip-rap of yesteryear [6].

Although the *Plan for Flood Control* was far from fully realized, the TRCA has achieved an impressive combination of infrastructure, land acquisition, research and awareness that has proved its worth on many occasions. The TRCA's Don Haley

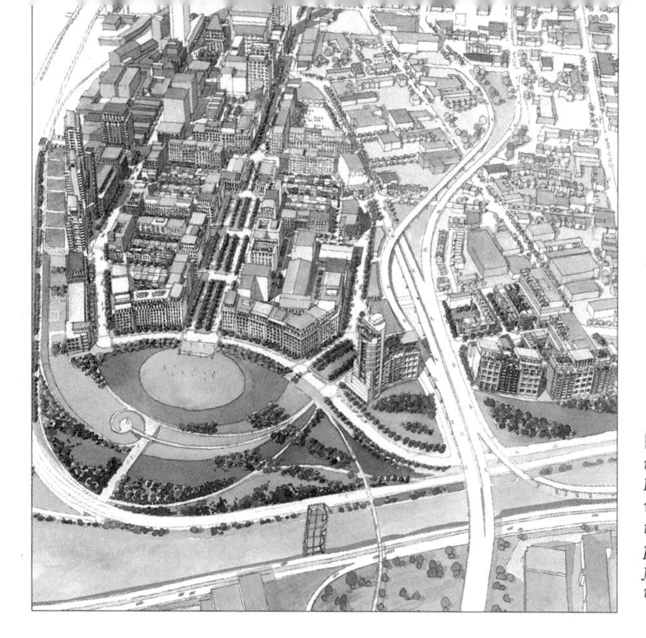

6 *Waterfront Toronto's rendering of the new West Don Lands community, with Don River Park in the foreground, 2007. The park will be built atop a flood protection landform, now under construction.*

explains that, over the past fifty years, they've done a good job managing flood risk from rivers and lakes. But today, and in the future, the more important concerns are what he calls urban flooding – on streets and in basements – from more frequent and more intense storms and the continued increase in urban density. Flooding that occurs in these areas – seen in events like Black Creek's August 2005 washout of Finch Avenue West – illustrate how critical it is to build resilient communities and ensure infrastructure is properly protected.

At the same time, Haley says, continued urbanization means that waterways and watersheds are always changing, and that organizations like the TRCA have to update hydrological data and flood-plain mapping more frequently to keep information as current as possible. Thankfully, he explains, the TRCA is not in this alone. With a number of public agencies and grassroots organizations tackling different issues and localities, the result amounts to a comprehensive flood-protection and water-conservation effort. While it may not be exactly what the 1959 *Plan for Flood Control* envisioned, the principles haven't changed.

Mahesh Patel

# The long haul: Integrating water, sewage, public health and city-building

In its early years, Toronto looked much like the major urban centres of today's developing countries. Squalor and overcrowding, paired with a lack of potable water and proper means of disposing of human and animal waste, made conditions ripe for the spread of diseases. In an editorial in the *Canadian Freeman* dated April 5, 1832, journalist Francis Collins described the unsanitary state of York, and what he thought should be done to remedy it:

> All the filth of the town – dead horses, dogs, cats, manure, etc [is] heaped up together on the ice [of Toronto Bay], to drop down, in a few days, into the water which is used by almost all the inhabitants of the Bay shore . . . There is not a drop of good well-water about the Market-square, and the people are obliged to use the Bay water however rotten. Instead therefore of corrupting the present bad supply, we think the authorities ought rather adopt measures to supply the town from the pure fountain that springs from the Spadina and Davenport Hill . . . There is nothing more conducive to health than good water – nothing more destructive than bad – and what ought the authorities to watch over and protect before the health of the community?

At the time, seepage from privies and the dumping of sewage into Lake Ontario contaminated drinking-water sources and caused outbreaks of diseases such as cholera, diphtheria and typhoid.[1] Over the following eight decades, conditions such as these drove authorities to examine the link between water, sewage and public health, and this, together with their willingness to resolve the conundrum, allowed Toronto to grow into a thriving metropolis.

At its inception in 1793, the Town of York was little more than a military camp established on the shores of Lake Ontario. It wasn't until three years later, when it became clear that York would become the capital of Upper Canada, that the site started to grow. By the early nineteenth century York's population had grown to 400 people and seventy-five houses. By 1832, it was larger than Kingston, its chief rival settlement. By 1834,

1 Typhoid is an infectious fever disease, often fatal, characterized by intestinal inflamation and ulceration.

the population had increased to 9,254, and York was incorporated as the City of Toronto. Much of York's rapid growth was attributed to immigration from Britain as well as its ability to exploit its capital status, including developing a port to foster commerce.

As the population increased and became more concentrated, basic human needs, including the availability of safe drinking water and the proper disposal of waste and sewage, became increasingly critical. The population began to experience the discomfort resulting from living a rural lifestyle in an increasingly urban environment. Residents relied primarily on wells, streams and water carters for their drinking water. These individualistic, private solutions to water supply were finally challenged in 1823, when York tendered its first public well for the town market square [1].

But the deteriorating, unsanitary conditions described by Francis Collins extended well beyond water supply. York householders used a cesspool, backhouse or pit privy – basically a large hole in the backyard – to dispose of their bodily wastes. Public sanitation consisted of similar holes called 'collects.' These 'collects' and box sewers also received dead dogs, cats and horses and other organic material. Fecal matter from overflowing pits and cesspools, along with material dumped by scavengers who cleaned the privy vaults, eventually ended up in nearby watercourses and Toronto Bay. It seems unavoidable that, with deteriorating sanitary conditions and the close proximity of wells to privies, both groundwater and bay water became polluted. Since the prevalence of unsanitary conditions and the dangers of sewage-contaminated drinking water were not yet recognized, it was only a matter of time before the town suffered its first public-health crisis. In 1832, cholera claimed over 200 lives in Toronto – including Collins' – and threatened its economy. The menace of economic downturn caused by the avoidance of the town by panicked outsiders prompted public action. Inspectors and superintendents appointed by the Magistrates of the Court of Quarter Sessions ordered wooden buildings and privies to be washed, inside and out, with lime. In the filthiest parts of the town, inspectors also ordered the daily collection of garbage and sewage and the building of box drains.

York's cholera outbreak also led to the creation of the provincial Board of Health, formed to guard against infectious diseases. The newly incorporated City of Toronto also instituted

1 *York's public well in the town market at King and Jarvis, before 1831.*

governance and administrative structures better suited to tackle the challenges of a growing city. The local Board of Health and the Board of Sewers were among the first administrative units to be created. However, although they recognized that unsanitary conditions contributed to the emergence of the epidemic, the understanding of the link between sanitary conditions and disease causation had still not been fully mapped out. Nevertheless, the cholera outbreak exposed the economic vulnerability of the city and put sanitation on the local political agenda.

Initially, the two main drivers of sewer building were the economic importance of keeping the city streets drained and the belief that the stench or miasmas resulting from filth and stagnant pools of water caused ill health. Evolving bureaucratic processes enabled the public to petition and inform Council of the need for new drains. In 1835, with the approval of the Royal Engineers, the building of Toronto's first six brick sewers, along King Street, began. The project incorporated a precedent-setting financing process that required property owners to make a one-time contribution to the cost of the proposed sewer.

Council's dilemma was that sewer building was capital-intensive and did not generate any profit. Once built, it was difficult to quantify the sewer's use. The lack of profitability meant Council could not attract private enterprises into developing and operating the sewer system. This required an innovative method of financing that would generate a continuous source of income to pay for local infrastructure improvements and allow the City to pay back loans in the form of debentures. That mechanism came in 1859 with the passing of an amendment to the Municipal Act that allowed local government to levy a tax on property owners, a precursor to the modern property-taxation system employed by local governments throughout Ontario. This legislation was arguably the cornerstone of Toronto's development into a major metropolis.

During the construction of the King Street sewer system, engineers made one critical change to the location of the sewer outfall, a change that would impact both commerce and public health. The sewers were originally intended to drain east into the Don River, but for some unspecified reason, the outfall was redirected south into Toronto Bay, at the foot of Peter Street.

By the second half of the nineteenth century, Toronto's economy had further expanded, its population had increased to well over fifty thousand people and conditions at the harbour continued to worsen as a result of the unresolved sewage-disposal problem. By the late 1860s, sewage accumulation in the harbour was disrupting the effective movement of cargo. Sewage plugged the outfall pipes, forcing water to back up into commercial buildings on the waterfront. And public health consequences were becoming more dire.

*[2] Distributing drinking water from City carts during an epidemic, September 1895. Private water carters likely operated in a similar manner in earlier decades.*

The matter of water supply was tackled first. In the 1830s, Toronto was a relatively young city with many pressing needs, street lighting and water for firefighting among them. The availability of water for domestic use by residents was not a priority, mainly because it could be secured from private and public wells and from carters who supplied water from other sources [2]. In 1840, local merchants petitioned Council to provide gas works for lighting, so the Standing Committee on Gas and Water Works consulted with the Montreal Gas Light Company, which culminated in Albert Furniss, a partner in the company, offering to provide Toronto with a gas works and lighting on a monopoly basis. His scheme accepted, Furniss shrewdly offered to supply water for the fire hydrants at a cost no higher than that he charged Montreal, as long as Toronto guaranteed him a twenty-one-year contract. Seeing an opportunity to establish a waterworks in Toronto, the committee convinced local insurance companies to finance half the city's commitment for the new system for five years. This deal neutralized any opposition to the new expenditure in Council, which accepted Furniss's proposal. By 1843, Toronto had its first waterworks system: a pumping station at the foot of Peter Street that drew water through bored wooden logs from the nearby harbour. But as soon as the waterworks system was operational, complaints emerged about its inadequate supply and the poor quality of the water it provided. Though disputes between the City and Furniss resulted in several ownership changes, he eventually regained controlled of the system.

This turmoil no doubt contributed to the lack of progress in the laying of pipes for potable water. Furniss's frugality and business acumen enabled his company to make a small profit during this period while avoiding any significant capital investments. Toronto's piped-water uptake rate was lower, however, than that of other North American cities, mainly because residents had ready access to other water sources such as wells and carters, which were better options than poor-quality piped water. And since it could not compel citizens to subscribe to water service, the company was not able to turn enough profit to finance capital improvements to the system. After Furniss died in 1872, the City purchased the utility from his family and created the Toronto Water Works Commission with the aim to completely rebuild and extend the system. This was largely achieved by 1877, just as the city's population topped the 70,000 mark.

As the population grew, so did the amount of raw sewage flooding into the harbour. During the Furniss era, the water-intake pipe adjacent to the Peter Street outfall had delivered polluted water to people's pots and cups. Toronto was afflicted with outbreaks of typhoid fever in 1845 and 1847 and cholera in 1849, 1854 and 1866. In an effort to remedy this predicament, the Water Works Commission moved the intake pipe to the lake side of Toronto Island after 1872. Water was piped through a 10,000-foot wooden conduit that ran through trenches and under Blockhouse Bay, and connected with iron pipes near the Western Channel, which brought the water to the mainland. The intent of this move was to use the island's sands as a gigantic natural filter. In 1881, the City also enacted a bylaw that required citizens to abandon private wells and take up the piped public water supply. Water-quality problems continued, however, as the island filter

[3] *Intake screen for a six-foot steel conduit, c. 1897.*

basin became increasingly putrid and weedy. In 1882, the *Globe* called Toronto's water supply 'drinkable sewage.' Public criticism resulted in the City moving the intake pipe further into Lake Ontario by 2,628 feet, and to a much greater depth – an approach that would be repeated over the next thirty years [3].

Meanwhile, in Great Britain, Edwin Chadwick's influence on public health practice was by now well-established. A proponent of the miasmatic theory, which focused on cleaning up the urban environment, Chadwick believed in the provision of a pure water supply and ending public dependence on privies and cesspools for human-waste disposal. It wasn't until the early 1880s that Chadwick's views became more widely accepted in Toronto, however. In 1883 the City of Toronto appointed Dr. William Canniff as the city's first permanent Medical Officer of Health. Canniff adopted a number of public health initiatives that Chadwick had proposed in Great Britain, including house-to-house inspections and the disinfection of privies, in an effort to raise awareness of the unsanitary conditions in the city. In 1884, Council enacted a new bylaw that gave the Medical Officer of Health the power to assess petitions for sewers and recommend their passage on sanitary grounds, where previously only the City Engineer had been allowed to assess these petitions. This measure later enabled Caniff's successors to abolish privies altogether and compel citizens to connect to the public sewer.

The 1891 census found 181,220 people living in Toronto. In 1893, the water-supply conduit under Blockhouse Bay ruptured,

[4] *The broken water-supply conduit in Toronto Bay, 1893.*

allowing sewage-contaminated harbour water to enter the piped-water system, resulting in a typhoid epidemic [4]. This event finally crystallized the connection between pure water, sewage disposal and disease. But cost considerations during this economically depressed time again led to a long period of inaction. Construction of a new drinking-water system, including the extension of the intake pipes, the construction of the Island Filtration Plant, a new 5,000-foot tunnel under the bay and an enlarged reservoir facility eventually began in 1908 [5]. But in 1910, technological failure struck again: broken tunnel equipment took in harbour water, contaminating the drinking supply and resulting in another typhoid outbreak. Luckily, this would be Toronto's last brush with this disease.

By the late nineteenth century, Louis Pasteur's germ theory of disease was well-established. At this time, Toronto, faced with various public health crises, was compelled to actively seek innovative solutions. The City successfully experimented with and adopted chlorine disinfection as a sanitization method, making Toronto a North American leader in the use of this disinfection technology. In 1912, a new and considerably improved intake pipe was installed at a depth of 100 feet, and work continued on upgrading the Island plant's filtering technology. These improvements, particularly chlorination, had a dramatic impact on typhoid death rates in Toronto, which dropped from over forty per 100,000 population in 1910 to fewer than two per 100,000 in 1915 [6]. By the latter date, City Engineer Charles

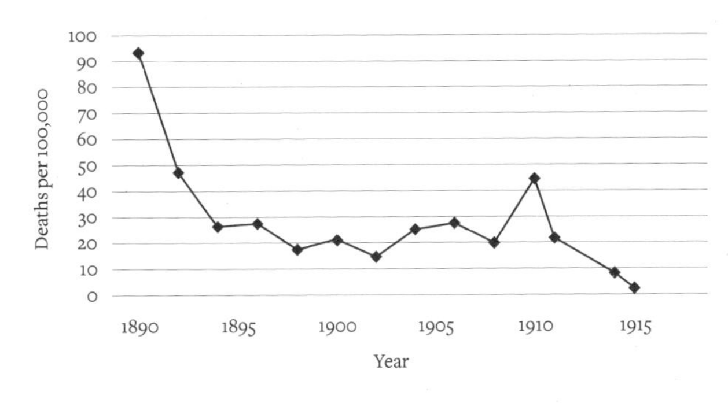


5 ABOVE *The new Island Filtration Plant nearing completion, c. 1913.*

6 LEFT *Typhoid mortality rates in Toronto, 1890–1915.*

Rust had been fired and replaced by a new Commissioner of Works, R. C. Harris, who would further overhaul Toronto's water-supply system.

While Toronto's water-supply predicament was being resolved, however, sewage disposal continued to pose considerable challenges. Public health advocates and engineers argued that a large-scale interceptor trunk sewer and disposal system was the only solution to the problem. The notion of intercepting and redirecting sewage before it reached the harbour had first appeared in 1850, but it was another twenty years before Council began to take greater interest in the problem.

For decades, civic action surrounding sewage disposal had consisted of professional report-writing and political exhortation. One proposal from the 1870s called for a large trunk sewer that would transport sewage eastward, discharging it into the Don River and eventually into the lake, but the $190,000 price tag was deemed too steep. In his 1883 inauguration speech, Mayor Arthur Radcliffe Boswell noted that '[the] Bay will soon become a cess-pool, and we cannot expect Toronto to retain the character for healthfulness it has hitherto borne ... See to it, then, that this great work be no longer delayed.' However, Council did not share Boswell's sense of urgency. The following year, the new mayor, Alexander Manning, addressed the same issues and triggered another round of reports. One report proposed two options: either transport the city's sewage eastward through a trunk sewer for open-water disposal, or undertake sewage treatment in conjunction with a trunk system. Manning was unable to convince the ratepayers (who then voted on every capital project) to approve the latter plan due to its $1.4 million cost. Two resubmissions and a memorandum to Council from Dr. Canniff, emphasizing the great benefits to public health of such a system, similarly failed. Council set up a Trunk Sewer Committee and called for more reports, even though an 1889 report by American consultants Rudolph Hering and Samuel M. Gray formally recognized sewage disposal and drinking-water quality as a joint problem. By the turn of the century, however, economic depression forced the City to suspend the matter from further consideration.

As the economy began to recover eight years later, Mayor Emerson Coatsworth forced the issue yet again, this time with the help of City Engineer Rust and a consultant from England, G. R. Strachan. The mayor's submission to the eligible electorate was rejected yet again in 1907. Finally, in 1908, upon the advice of the provincial and local boards of health and of the Faculty of Medicine at the University of Toronto, Toronto's ratepayers finally approved the construction of a relatively sophisticated $2.4 million sewage treatment system. It included an intercepting sewer, pumping and chlorination stations, a primary treatment plant at Coxwell and Eastern avenues and a lake outfall southeast of Ashbridge's Bay. The system was completed in 1913. The site remains operational to this day, although it is now overshadowed by the massive Ashbridges Bay Wastewater Treatment Plant immediately to the south [7].

Many circumstances conspired in the transformation of York, the military camp, into Toronto, the major Canadian

metropolitan centre. One significant factor was the ability of its citizens to make the link between public health, water and sewage under conflicting and sometimes perplexing circumstances. Without the benefits of hindsight and today's know-how, the city fathers saw the value of investing their limited resources for the greater good. They benefited their generation and generations to come by opting for publicly owned sewer and water works, having recognized that, for private enterprise, public health would always be secondary to profit. Clearly, improved water quality resulted in a healthy population, and the ability to provide piped potable water and remove and treat sewage enabled Toronto to grow. Local financing issues and the broader economic climate often delayed critical infrastructure investment. Historically, this was a source of much frustration, and it remains a problem in today's Toronto. This begs the same sort of question that Francis Collins asked in 1832: 'What value do we place on the health of the future residents of our city?'

7 *The new Main Disposal Works under construction near Coxwell and Eastern avenues, June 20, 1912. Greatly enlarged, this building continues to act as a wastewater pumping station.*

Steven Mannell

# A civic vision for water supply: The Toronto Water Works Extension Project

In February 1928, fifteen years into his tenure as Toronto's Commissioner of Works, Roland Caldwell Harris finally held in his hands a set of preliminary design drawings for the filtration plant at Victoria Park, the first component of the Toronto Water Works Extension (TWWE) project. Set out in a report he issued in 1913, then delayed by the outbreak of the Great War and the conservatism of Toronto's Board of Control, the project was revived and expanded in a 1926 consultant's report, and embodied Harris's civic vision of a water-supply system worthy of Toronto's future, set within beautiful public parks and linked to a grand system of boulevards and bridges encircling the city.

R. C. Harris was a career bureaucrat at a time when the word *bureaucrat* was a term of approval and status, used to describe professional managers and facilitators. Progressive-era magazines like *The American City* supported the self-awareness of a new class of knowledge workers, the city managers whose terms of office were not tied to the patronage of elected officials or behind-the-scenes fixers and bosses. Bureaucrats and magazines shared a vision of new possibilities for cities in which public services and amenities would be the systematic and planned responsibility of city government, not the capricious spoils of ward-heeling and patronage.

Harris came into this milieu as the proverbial golden boy of City Hall. He began his municipal career as a clerk in Toronto's Property Department and worked his way up to become Property Commissioner at the age of thirty. His identification with civic work was total; from 1902 until 1915, he lived with his family in a third-floor apartment at city hall. In 1912, Harris was named Commissioner of Works and City Engineer despite having no engineering credentials. His administrative drive was well-recognized, however; in announcing Harris's appointment, Mayor Horatio Hocken noted his 'aggressive militant earnestness.'[1]

Once installed, Harris set to work on the Prince Edward Viaduct, which would connect Bloor Street to Danforth Avenue over the Don Valley. It was the first visionary scheme of the new century, one that relied on a Works Department–directed design and considerable input from the commissioner.

1 Quoted in *Architecture of Public Works R. C. Harris Commissioner 1912–1945* (Toronto: Market Gallery of the City of Toronto Archives, 1982).

Completed in 1919, the viaduct would be a prototype for the many bridges built by the Works Department over the next decades, with its clear articulation of the steel substructure from the concrete roadway and balustrade, and the integration of roadway and walkways, sewer and water mains and provision for public transit.

Water supply was a hot topic in progressive city-management circles at that time, and failures in Toronto's water-supply system had been the instrument of Harris's ascendance to the Works Commissioner's office. Water may have taken on a personal dimension for him as well. In a winter 2006 *Spacing* magazine article, John Lorinc speculates that the death of Harris's infant son, Emerson, in 1906 was the result of a water-borne disease. Whatever the specific mix of motivations, Harris moved quickly to report on the local supply crisis.

In his 1913 TWWE project report, Harris proposed a new waterworks at Victoria Park, just beyond the eastern city limit. In system terms, the proposal was a singular addition to the water system, really just another source and aqueduct. But the report was visionary in civic terms: 'It is proposed to erect handsome buildings [at Victoria Park], which, in conjunction with the park section and the beach, will constitute one of the most beautiful areas in Toronto.' This vision of a grand waterworks set in a public park near the Eastern Beaches south of Queen Street would remain a cornerstone of future plans for water supply and set the high civic tone to be matched in all later work.

Unfortunately, war, and the reluctance of Toronto's Board of Control to enter into debt financing of City projects, conspired to place this grand vision on hold. When Council voted in 1915 to repeal the 1913 Victoria Park expropriation bylaw, Harris moved his family to a house nearby, on Neville Park Boulevard, standing vigil over his project. Expropriation was finally carried out in 1923, and H. G. Acres and William Gore were commissioned to report on a duplicate waterworks for Toronto. Acres was renowned for his work on the Niagara hydroelectric installations, while Gore was a key figure in the development of water-filtration technologies. In Gore's office worked his partners William Storrie, a design engineer, and Colonel George G. Nasmith, a bacteriologist who had ensured the purity of water supply to Canadian troops on the Western Front.

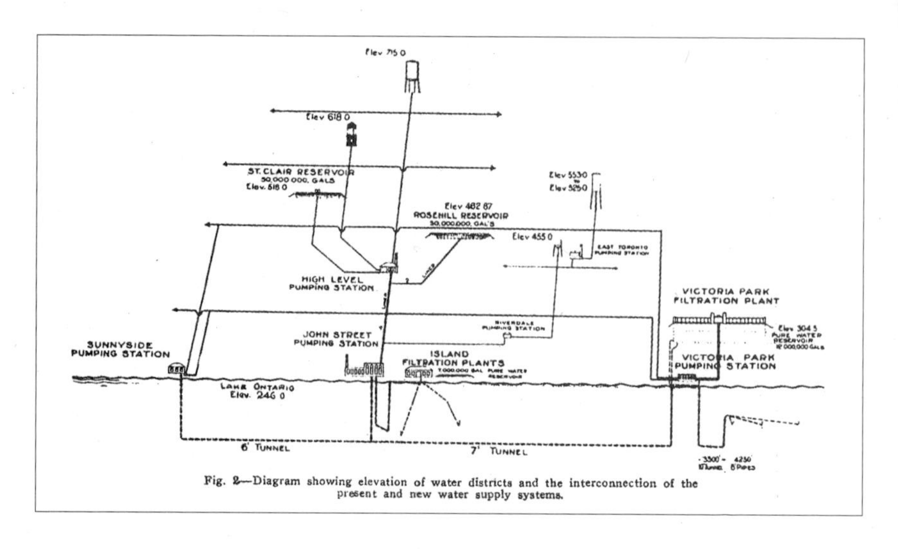

Fig. 2—Diagram showing elevation of water districts and the interconnection of the present and new water supply systems.

**1** *Elevation schematic of the Toronto Water Works Extension project as proposed in the Acres and Gore Report of 1926.*

The 1926 Acres and Gore Report reads as a new testament to the old testament of Harris's 1913 report. A fully duplicate water-supply system is recommended, which would have doubled the city's filtration and pumping capacity and provided multiple water-distribution paths [1]. Acres and Gore added several new components to Harris's planned Victoria Park filtration-plant and pumping-station designs, including a reservoir and tank at St. Clair and Spadina near the existing High Level Pumping Station; a filtered water tunnel across the lakefront from Victoria Park to the existing John Street pumps and a new pumping station at Parkdale; and a connection to the existing Island Filtration Plant. Beyond the doubled filtration capacity, this additional reservoir capacity and the distribution of pumping throughout the system allowed for the full redundancy of water supply and distribution, protecting citizens from water shortages and ending the city's vulnerability to complete interruption of service due to the failure of any single element. As a whole, the Toronto Water Works Extension imposed a comprehensive network of water supply onto the entire Toronto landscape, and sustained Harris's civic vision. Each element was placed in proximity to other new civic features: at St. Clair Avenue, to the proposed boulevard with its dedicated streetcar allowance; at Parkdale, to the recently completed Lake Shore Boulevard, with its public beach, bathing pavilion and amusement area.

Detailed design work began in earnest in 1927 in the engineering offices of H. G. Acres & Co. of Niagara Falls and Gore,

Nasmith & Storrie of Toronto. Harris insisted on being briefed throughout every stage of the process. His meticulous review of the preliminary drawings for the filtration plant at Victoria Park, sent to him in February 1928, provoked a strong letter to the engineers criticizing the quality of the architectural expression. Written fifteen years after his initial report, following many delays and frustrations, Harris's letter to William Storrie expressed not relief or satisfaction that the real work had finally begun, but rather a restless concern that the realization would not measure up to the vision. Harris was blunt and unequivocal: 'The buildings as shown on the perspective sketch appear to me to be plain and unattractive.'

Storrie's reply, sent two days later, is backpedalling and contrite. Over the next months, Gore, Nasmith & Storrie's staff architect, Thomas Canfield Pomphrey, spent substantial time on the project that culminated in a large ink-and-wash rendering of the filter/administration building completed in early 1929 [2]. This drawing set the architectural vision for the TWWE on par with Harris's civic ambitions, and placed Pomphrey in a position of remarkable authority and influence with the commissioner.

THE PALACE OF PURIFICATION T. C. Pomphrey had apprenticed as an architect in Scotland prior to immigrating to Toronto in 1906, where he worked for leading architects including John Lyle and Darling & Pearson. He was severely wounded in the right shoulder in the war and spent a number of years in convalescent hospitals before resuming his practice in Toronto. In 1920, Pomphrey joined the firm of Gore, Nasmith & Storrie as staff architect and, in 1924, he and architect William Ferguson won the 1924 competition for the Great War Cenotaph in front of what is now Old City Hall.

Pomphrey's 1929 rendering of the Victoria Park filter/administration building represents a fundamental transformation of the character of the waterworks from the scheme shown in the 1926 Acres and Gore Report, which had reserved architectural expression for the 'head' administration/laboratory building while plain filter galleries trailed behind in a supporting role [3 4]. Pomphrey introduced a cross-axial basilica plan, familiar from churches, that monumentalized all parts of the works. Administration and labs occupied a central north/south axis between two wings of an east/west-oriented filter gallery.

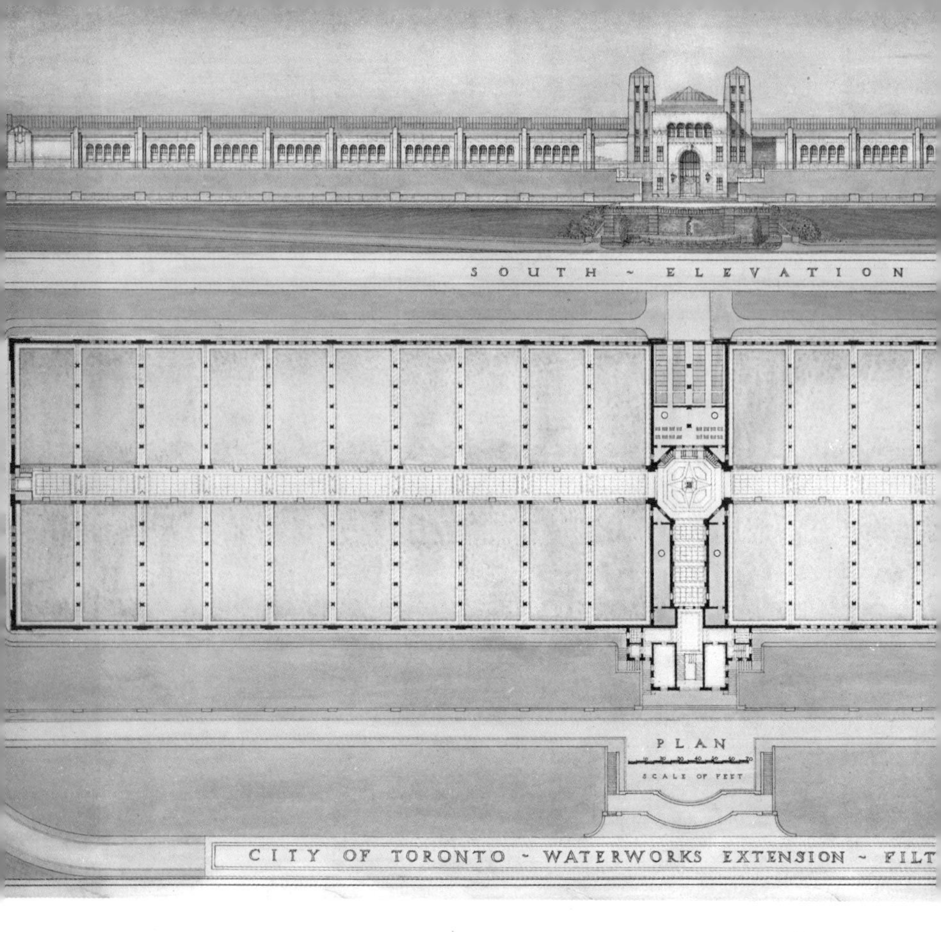


2 *Thomas C. Pomphrey's ink-and-wash rendering of the south elevation and ground-floor plan of the Victoria Park filter/administration building, 1929.*

A central rotunda with a monumental signal pylon occupied the crossing of axes, while the 'sanctuary' at the north end of the axis was reserved for the chlorinators, which were visible behind a glass wall, symbols of Toronto's proud contribution to the water-supply process.

Pomphrey's basilica stretched nearly the full width of the site, oriented to present a monumental face toward the view from the lake and lakefront, stepping up and back on a series of terraced landforms arranged around a major axis defined by the intersection of the intake tunnel with the seawall. Elaborate driveways, walkways and stairs brought the visitor around to the centre of the south front, which culminated in a terrace with views onto Lake Ontario, the source of raw water, and was backed by an entrance in the form of a triumphal arch flanked by towers. Terraces, retaining walls and crisp geometric slopes of grass provided a podium for the basilica, and a

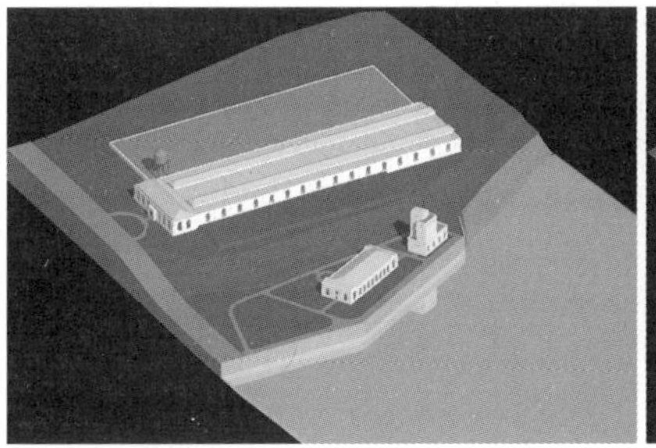

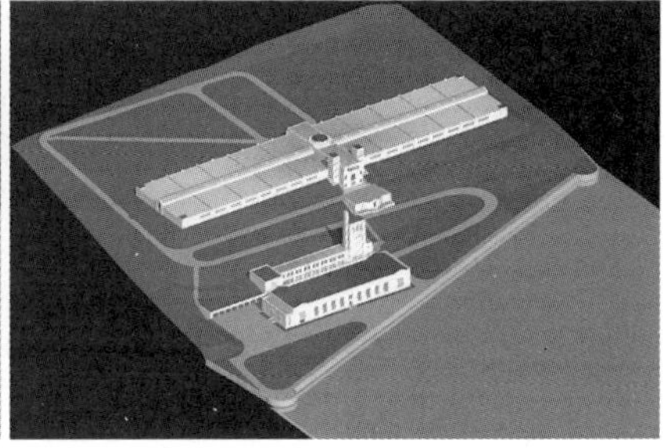

[3 4] *Development of the Victoria Park Water Works design, as envisioned in the 1926 Acres and Gore Report (left) and Pomphrey's redesign of 1928–1929 (right).*

backdrop for the pumping and service buildings downslope toward the shore.

The plant's intake, pumping and filtration processes followed the ordering principles of the architecture in both plan and section. Raw lake water entered the site on the central axis, moving laterally as it passed through various process stages but always returning to the axis before moving on to the next step, culminating in a return to the central rotunda for chlorination before leaving the site in the pure water mains. Not just a matter of adding a 'skin' of architecture to an engineering carcass, Pomphrey's architecture reconfigured the entire site and water-purification process along Beaux-Arts architectural principles of symmetry, proportion and eurhythmy. The waterworks were designed to accommodate the public for tours, and to remind plant workers of the public nature of their work. High-quality materials sustained the tone, both inside and out: yellow brick and Queenston limestone; copper flashings and roofs; terrazzo floors, herringbone tile wainscoting, black marble and limestone trim and bronze hardware. Details provided an iconography of water process, especially the carved turbine frieze at the top of the pumphouse walls, and the stylized 'TWW' emblems. Controls and monitors were enlarged to monumental scale. Pride of place in the filter building's rotunda was reserved for the signal pylon, whose lights and dials make filtration and water-supply quantities visible and symbolic [5 6].

Numerous articles on the Toronto Water Works Extension were published in engineering journals in the late 1920s and 1930s, with water filtration, pumping and distribution being the principal focus. But a large reproduction of the rendering accompanies all these articles, declaring unequivocally the central role of monumental architectural expression in the project. Harris's assessment of Pomphrey's role is expressed in a 1933 memo in which Pomphrey is authorized to make decisions

[5] [6] *Interior views of the completed filter/administration building at Victoria Park, January 18, 1935. The west filter gallery looking toward the central rotunda (left), and (right) the filter rotunda showing the signal pylon that broadcasts information about operations.*

on-site regarding architectural details, materials and specifications – a remarkable delegation of power that Harris did not assign to Gore or Storrie, the principal consulting engineers and Pomphrey's employers.

The key published photographs of the Victoria Park plant, which was completed in 1938, were taken from offshore on Lake Ontario, reflecting the conscious urban mythmaking that underpinned the project [7]. Two purely symbolic elements punctuated the ensemble: the alum tower, topped by a mysterious belvedere looking out over the lake, and the terrace building, whose niche and bronze fountain marked the line of the incoming raw-water conduits buried below. The waterworks projected a new face of the city outward, an embodiment of a mature, self-conscious sense of the civic.

THE TOWER OF PURE WATER Seven miles northwest of Victoria Park, the St. Clair Reservoir was the first completed element of the TWWE project, comprising a covered reservoir topped by a public park that overlooked the wild Nordheimer Ravine landscape. The complex was organized in relation to the view from the monumental new bridge that carried Spadina

7 *The 'Palace of Purification' (the Victoria Park Water Works, now the R. C. Harris Water Treatment Plant) seen from offshore in Lake Ontario, December 17, 1936. In the foreground, just behind the seawall, is the temple-like pavilion of the pumping station; behind its east facade is the alum tower, topped by a belvedere. A walkway and stair climbs the slope on the central axis of symmetry, leading to the terrace with its niche and bronze fountain. Behind the terrace is the triumphal arch housing the main entrance to the filter/administration building, flanked by towers. The filter galleries stretch away to the west; the eastern galleries would not be built until the 1950s.*

Road to its intersection with the proposed civic boulevard of St. Clair Avenue. For Harris, the St. Clair site offered an opportunity to bring together in one project his obsessions with waterworks, bridges, boulevards and public parks. The character of the park was affirmed in a letter from Gore to Harris in early 1926: 'The whole will be sodded over and sidewalks and flower pots could be placed on same.' As described in a *Canadian Engineer* article by Gore from 1929, the civic aims of the design were more ambitious: 'The reservoir has been made to conform to extensive improvements in roads and bridges in the vicinity, and the grounds are to be laid out and planted for use as a public park.'

Road bridges were as prominent as waterworks in Harris's vision of a renewed Toronto. By the late 1920s, the Works Department's designers had developed a characteristic bridge design based loosely on the Prince Edward Viaduct, with a steel truss substructure surmounted by a concrete road deck, pylons and rails. Deployed in various situations around the city between 1923 and 1939, the consistency of the language of those bridges reinforced the systematic nature of public works in Harris's vision of the city.[2] At the St. Clair site, City engineers designed a replacement for the old wooden trestle bridge on Spadina Road at the same time as they produced the reservoir design process. Completed in 1931, the new bridge provided an appropriate frame and viewing platform for the reservoir and park, along with a grand approach to the even grander St. Clair Avenue.

Seen from the bridge, the site planning of the St. Clair reservoir included a terraced landscape that rose from Nordheimer

2 The St. Clair Avenue viaduct over the Vale of Avoca of 1922–1924 and the Mount Pleasant Road bridge over the Blythwood Park Ravine of 1934–1937 are examples.

Ravine, and two symmetrically arranged reservoir chambers that flanked two neoclassical pavilions marking the top (valve house) and bottom (pipe tunnel portal) of a monumental stair built atop the buried water mains. Ornament included generalized neoclassical devices such as urns, deep rustication and prominent pediments. Construction of the reservoir and its landscape was completed in 1931.

The plans also proposed that an ornamental overhead tank be situated on the triangular earthen prow at the south end of the reservoir, a tower that would command views from both the new bridge and the ravine below. The tank's function was to provide thirty minutes of water pressure and supply to a small pumping district (District No. 3) that lay above the reservoir supply elevation in the event of a power failure to the electric pumps. When reservoir construction began in 1929, Harris became anxious to view the tank design, hectoring Storrie with requests for Pomphrey's sketches. Once the design was finalized in 1932, a line drawing of the tower elevation took the place of the Victoria Park rendering in trade-journal articles on the TWWE, the new exemplar of architectural quality and ambition [8 9].

These drawings show a standard steel water tank, encased by vertical limestone pilasters framing panels of yellow brick. A rusticated limestone portal around the service door would have led to a roof-access ladder housed within a projecting bay surmounted by a small dome and flagpole. Copper spandrel panels bearing the 'TWW' device were to be set between the pilasters, the strong blue-green vertical panels of copper vividly signalling the reserve of pure water to be stored within the tower.

Construction of the ornamental overhead tank was highly controversial in 1933, with Toronto's Board of Control unwilling to commit $70,000 for the ornamental cladding on top of the $43,000 cost of the inner steel water tank. The *Toronto Star* felt that the cladding would be 'a credit to the city and to the fine district in which it will be erected,' but at a time of economic difficulty, many felt the 'fine district' alone should bear the added cost. In September 1933, the Board of Control formally rejected all tenders for the tank. Drawings were prepared for an unadorned version, but Harris instead chose to place the construction of both the steel tank and the ornamental cladding on hold in hopes of a more sympathetic future political climate.

The deferral of the St. Clair tank was a harbinger of the controversy that would surround the completion of the TWWE project.

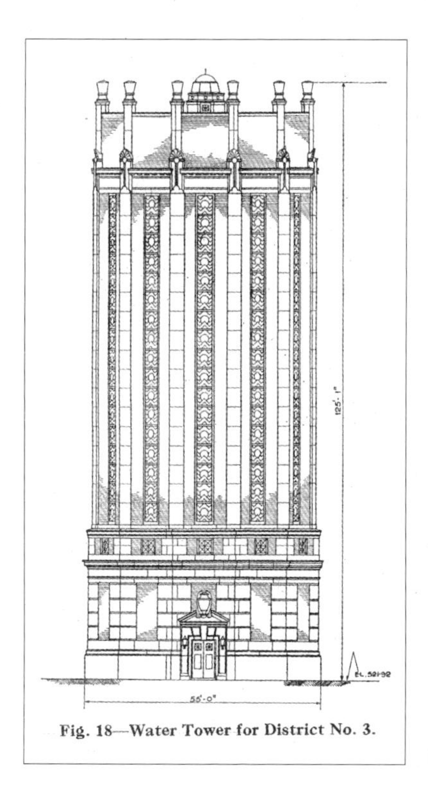

Fig. 18—Water Tower for District No. 3.

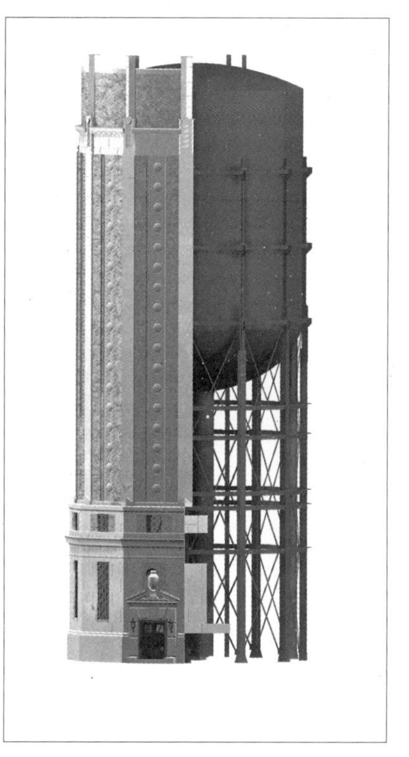

[8] LEFT *Line drawing of the front elevation of the District No. 3 overhead tank proposed for the St. Clair Reservoir, designed by Thomas C. Pomphrey, 1932.*

[9] RIGHT *Digital rendering of the 'tower of pure water' (District No. 3 overhead tank) shows the architectural mask of stone, brick and copper panels partly stripped away to reveal the steel water tank, 2000.*

As the Depression deepened, the high level of material and spatial quality became a political liability. Without ceremony, water began to flow through the Victoria Park plant on November 1, 1941. (It was given Harris's name after his death in 1945.) Completion of the TWWE continued in fragments; the Parkdale Pumping Station and Surge Tower were built in the early 1950s on a workmanlike site plan in minimally detailed red brick.

As for the St. Clair tank, the foundations constructed under the 1929 reservoir contract were maintained diligently, protected by a timber hoarding. In 1947, the Ward 4 Committee of Women Voters complained to Harris's successor, G. G. Powell, of the unsafe and unsightly condition of these foundations, and demanded that they be removed. Powell instead ordered the foundations to be protected by new hoardings. Harris's dream for the tower of pure water remained alive [10]. But, by the 1950s, improvements in electric service had eliminated the risk of power failures that the tank was meant to guard against. Today only a mysterious ring of concrete remains at the southern prow of the reservoir: the top of the foundation wall.

[10] *A digital rendering of what could have been. The unbuilt District No. 3 overhead tank in context of the park and playing fields above the St. Clair Reservoir, viewed from the corner of St. Clair Avenue and Spadina Road, 1934.*

The ultimate moment of truth for the R. C. Harris Water Treatment Plant came in 1955. Metropolitan demand for water had grown sufficient to justify the completion of the east wing of the filter building, enabling the plant to produce a total of 200 million gallons of pure water per day. What Commissioner Harris understood as 'public works,' which coupled function with a visible civic role, was by this time increasingly called 'infrastructure,' with its invisible omnipresence. After much debate, Works Commissioner Leslie Allan finally came down on the side of public works, approving the use of decorative stone in the new wing. The east wing was carefully detailed in accordance with Pomphrey's original design and contract drawings. This was much discussed in the engineering and utilities trade press. In the post-war climate of progress and utility, architectural treatment of public works had again become remarkable [11].

Harris imagined a civic network with interlinked water and sewage systems, and also streetcars, boulevards, parkway drives and bridges. He thought the city should provide both service and amenity in monumental forms as a means of fostering civic spirit and a willingness on the part of citizens to pay taxes and fees, and to encourage city politicians to support debenture issues. Acres, Gore and Nasmith imagined a public health and engineering network. Pomphrey imagined orderly form, the symmetry and eurhythmy of parts and wholes. His architecture governs and idealizes the engineering process, and gives symbolic expression and tangible material value to the supply

of pure water. Consistent language of form and material relates each visible component to the others, making the entire project legible as a systematic ordering of the Toronto landscape. The intersection of the visions of these men is the Toronto Water Works Extension, a landscape of water that forms a lasting legacy to its city. That legacy remains well-tended. Over the past decade, careful stewardship by Harris Plant staff and manager Ron Brilliant (working with a dedicated Public Advisory Committee) has resulted in the incremental restoration of many original features, while a major new Residue Management Facility completed in 2008 was designed by Ken Mains of CH2M HILL (successor firm to Gore, Nasmith & Storrie) to be largely underground, avoiding conflict with the original design for Victoria Park.

*11 The 'Palace of Purification' completed. Aerial view of the R. C. Harris Water Treatment Plant after completion of the east wing of the filter building in the late 1950s, completing Pomphrey's 1929 composition.*

Nearly a century after it was first envisioned, the TWWE project remains sufficient to meet the now much larger city's water needs (excluding summer lawn watering); it maintains a strong record of public health, having eliminated cholera, typhoid and other water-borne menaces from the lives of citizens; and its major sites, including the Harris Plant and the St. Clair Reservoir, provide some of the most elegant recreation grounds in the city. The Toronto Water Works Extension continues to mark a decisive period of the twentieth century when civil engineering and public works were understood to be great civic deeds, demanding a great civic architecture.

Michael McMahon

# We all live downstream

> A birdwatching resident of the Rosehill district to the southeast of St. Clair and Yonge espied an engineer boring test holes recently around the open reservoir that gives the district a cool sparkling oasis in summers.
>
> Inquiries confirmed the suspicion that a plan was afoot to tamper with the Rosehill reservoir. Further investigation yielded the information that the Metropolitan Works Department, which has charge of the city's water supply, planned to put a lid on the neighborhood's private lake.
>
> A save-the-reservoir petition was circulated by residents, including Lady Roberts, wife of Sir Charles G. D. Roberts. It attracted 368 signatures, including those of Dr. Charles Best, the co-discoverer of insulin, and Dean Bladen, the renowned political economist.
>
> The upshot of all this has been a reassessing of what is to be done at Rosehill – and why.

So begins a *Toronto Telegram* editorial from June 15, 1961, that was clipped and circulated within the Metropolitan Toronto Works Department of the day. The story reveals some of the complexities of our urban environmental history. And here the possessive *our* needs to be read in the most intimate of senses: our bodies are that part of nature that we directly inhabit. There, perhaps more than in other parts of our individual and collective cosmologies, the water that sustains us bears history's burdens, terrors and wonders. After nearly a century of chlorination, we no longer need to worry about dying because of some waterborne germ. But better living through chemistry, seen in the context of where our water comes from and goes and the treatment steps in between means that, more than ever, we are likely to get cancer in ten or twenty years. The temporal dimensions of public and environmental health are changing, and we are left with two questions: 'What is to be done?' and 'Why?'

Urban infrastructure systems are often portrayed as schemes through which cities tap into and impact their surrounding natural environments. And in such portrayals, nature – God-given or otherwise – is often idealized as a pure and pristine baseline from which we might measure and even

improve on our fallen, urban circumstances. The idea of urban nature, by contrast, starts with the presumption that there are few, if any, givens. Urban environmental history as it began to be written in the last decades of the twentieth century highlights complex, long-standing entanglements between nature and society. More recently, urban political ecologists have begun to expand upon some of the activist energies described in the *Telegram* editorial. The story of the Rosehill Reservoir, and the role of citizen-led activism as a promoter of more sustainable technologies, are an example of why Toronto must pay more heed to its community activists if it is to truly claim to be a leading Great Lakes city.

*[1] Water intake being lowered into Lake Ontario off Victoria Park, June 19, 1933. But the water intakes and sewage outfalls existed in relatively close proximity to each other. By 1913 the search for pure and pristine water supplies was largely abandoned, having been replaced by visions of a variety of technical fixes. 'Upstream' of the intake pictured above, polluted lake waters were soon to be made safe for human consumption with the help of mechanical and chemical engineering at the new R. C. Harris Water Treatment Plant.*

### THE ROSEHILL RESERVOIR: FROM LAKE TO STORAGE FACILITY

The Rosehill neighbourhood has long been a well-to-do and well-appointed sort of place. The construction of the Rosehill Reservoir in the mid-1870s added another scenic element to the neighbourhood; residents were able to take in evening promenades around the reservoir's open waters on a footpath provided for that purpose, and could enjoy views to Lake Ontario – the source of the waters in the reservoir – from atop the ancient Lake Iroquois shoreline, where the reservoir was located. While residents admired the view, water, having been pumped up from the Great Lake [1], flowed back to it by way of gravity, delivering drinking water and carrying away household, commercial and industrial wastes.

A product of nineteenth-century civil engineering, the Rosehill Reservoir initially functioned as an organic machine. Water that was often polluted and even deadly in the 1880s and

*R. C. Harris Water Treatment Plant's chlorinators, August 5, 1955. Works Commissioner Harris was apparently very proud of chlorinators such as these. In the 1930s, they represented the ultimate in a chemically based, technical fix for Toronto's water.*

1890s was stored and cleansed in the reservoir with the help of ultraviolet rays from the sun and the interactions of wind, oxygen and micro-organisms on and in the reservoir itself. However, by the mid-twentieth century, such reservoirs had become redundant because they were deemed to be 'downstream' of water-processing facilities such as the R. C. Harris Water Treatment Plant. Maintaining an open reservoir of the sort that still exists in New York City's Central Park was considered by water bureaucrats to be uneconomical and unhealthy because water there would have to be chlorinated twice – once at R. C. Harris and once at Rosehill – to properly control for algae and counteract the presence of potentially dangerous micro-organisms [2]. The issue of cost, over and above any clear-cut issues of public health, appears to have been the main factor in the 1961 decision to transform the Rosehill Reservoir from an organic machine into one of a series of enclosed storage facilities for Metro Toronto's waterworks system.

At the time, Works Commissioner R. L. Clark stated categorically that the proposed million-dollar reservoir lid, 'capitalized over 20 years,'[1] would save money that would have been spent on the extra cleaning, chlorination and monitoring needed for an open reservoir. But, as it was reported in the *Toronto Star*, Rosehill's save-the-reservoir petitioners were 'unimpressed' with such arguments. G. T. Scroogie, deputing on behalf of residents at one Works Committee meeting, declared that 'the aesthetic appeal of this miniature lake in the center of the city should overrule considerations of economy or efficiency.'[2] We can only speculate on how residents reacted to Clark's concerns with 'steadily increasing amounts of fall-in from city air pollution.' The neighbours may have argued that municipal departments should work together to reverse air-pollution trends rather than use pollution as an excuse to re-engineer the local lake out of existence. Regardless, by 1967, the Rosehill Reservoir waters were completely enclosed, protected from fallout of any sort.

The Rosehill protests exemplify the citizen activism rampant in Toronto in the sixties and foreshadow more general public concerns with Spadina Expressway–scale infrastructure proposals. From the late 1950s to the early 1960s, anti-expressway protests arose to save elements of the city's built and natural heritage such as Fort York and parts of the ravine system. The actions of the Rosehill save-the-reservoir petitioners, however,

1 Statement of Commissioner Clark as summarized in editorial, *Toronto Telegram*, June 15, 1961.

2 'Residents Protest Covering Reservoir,' *Toronto Star*, May 27, 1961, p. 35.

were unique, concerned as they were with a very particular assemblage of biological, technical-chemical, landscape and built-heritage elements. Residents saw the Rosehill Reservoir as the biggest city 'lake' this side of Grenadier Pond and regarded it as a vital piece of metropolitan nature.

### LOOKING UPSTREAM FROM THE ROSEHILL RESERVOIR

The Rosehill Reservoir was built at a time when the City of Toronto was still undecided about its source of drinking water. In the 1880s, engineers proposed that Rosehill draw its supplies by gravity from headwater sources on the Oak Ridges Moraine. If this had happened, Toronto would have conformed to the nineteenth-century engineering ideal of a river city: one that takes pure and wholesome water from upstream and dumps its water-borne waste downstream. For reasons of economy, and because of the need for a larger and more flexible water supply in the face of growing demand, Toronto decided to stay with Lake Ontario as both its main source of water and the sink for its sewage. Outbreaks of water-borne typhoid fever persisted as a result, until the introduction of chlorine to Toronto's water supplies just prior to World War I.

Chlorine, however, can be both friend and foe. Since the 1980s, the public has begun to understand that the uncontrolled by-products of chlorine, so-called organochlorines, comprise a truly modern-day imbroglio. Because of the limited and controlled use of chlorine, we no longer die of typhoid. But the weight of evidence suggests that an organochlorine-saturated ecosystem substantially increases our chances of getting cancer. Those of us who are serviced by Toronto's municipal water system are safer now than we have ever been from death by tap water.[3] Yet new knowledge of chlorine chemistry allows us to see that public health problems were compounded at the Rosehill Reservoir for a time. From the mid-1920s, the City began to chlorinate the water twice to stave off the effects of what the civil engineers of the day dubbed 'progressive pollution.' The second time, at the reservoir, the dispensation of chlorine was not well-controlled [3]. Given the chlorine-based water and wastewater circulation system, the covering of the Rosehill Reservoir in the mid-1960s very likely had marginal public health benefits in the context of a larger sea of rising cancers. The 'downstream' side of Toronto's hydro-social cycle[4] provides various windows onto this sea.

3 In the mid-1990s, Toronto cancer data was not included in many of the epidemiological 'meta studies' concerned with generating the best weight of evidence of correlations between chlorinated water supplies and various cancers. My understanding for the reasons Toronto was off the radar of such studies is all for the good. Toronto has a long history of advancing chlorine chemistry, to the point that chlorine residuals needed to keep water pure in old piping are chemically stabilized so as to minimize organochlorine formations. I continue to drink and enjoy the taste of Toronto water, finding the bottled water alternatives abhorrent. But I wonder what the price tag of ozone alternatives to chlorine at the public scale might be for Toronto water. Such questions become significant in the context of calls for chlorine sunsets, as discussed on the following page.

4 Thanks to Roger Keil, director of the City Institute at York University, for this phrase. The notion of a hydro-social cycle is particularly apt for 'lake' cities such as Toronto, if not for somewhat more linear 'river' cities on the Great Lakes like Detroit, Buffalo and Chicago.

[3] *Chlorine-dispensing boat and apparatus, Rosehill Reservoir, c. 1955. Toronto's metabolism of water and waste continues to entail chlorination at both the intake and outfall ends of Toronto's hydro-social cycle. Chlorination at in-between points such as the Rosehill Reservoir, however, was abandoned in the mid-1960s. Retrospectively, we can say with considerable certainty that this was a good thing. Liquid chlorine, dispensed in the fashion shown above, very likely gave rise to a variety of carcinogenic organochlorines, even as it protected consumers from shorter-term illnesses and even death by water-borne germs.*

**WE ALL LIVE DOWNSTREAM: TORONTO AS A GREAT LAKES CITY?**
Like most cities on North America's inland seas, Toronto began to look to its Great Lake as the ultimate sink for its wastes from the late 1800s. This was a time when much older ideas about the self-purification of water came to be inflected with new germ theories. Germs and other micro-organisms, properly fed with oxygen, came to be seen to have both good and bad properties. Where human, water-borne wastes were concerned, the Great Lakes were deemed to have assimilative capacities, which, if surpassed in the near-shore regions of the city, could be complemented by large doses of chlorine, which would kill the stench of too-slowly-oxidizing feces [4]. However, continuing bad sights and smells and high E. coli levels indicate that assimilative capacities had been overtaken, especially after combined sewer overflow events on Toronto's shores. But in the era of the Canada-U.S. Great Lakes Water Quality Agreement, first signed in 1972, a new and more shocking realization began to take hold with regard to the 'physical, chemical and biological integrity of water,' or lack thereof. Organochlorines, in any quantity, came to be regarded as a serious danger, given their carcinogenic properties, and their environmental persistence and abilities to bioaccumulate and bio-magnify in living organisms. Ecologists like Barry Commoner drove such points home in widely circulated articles from the mid-1980s, which were echoed by Greenpeace's Water for Life tour of the Great Lakes in 1988 and protests at the stormy Great Lakes meetings of the Canada-U.S. International Joint Commission in Hamilton in 1989. This spate of activism had positive results: the IJC, which had been on the front lines of the 'virtual elimination' of toxic chemicals from the Great Lakes since 1978, supported calls for the phase-out of chlorine as a feedstock chemical in 1992. Since that time, however, Toronto has been throwing increasing volumes of chlorine at its wastewater problems. In 1993, the super-chlorination of combined sewage overflows at the Ashbridges Bay Wastewater Treatment Plant [5] gave rise to provincial charges of 'excessive chlorine toxicity . . . without any real disinfection being achieved.'[5] The Province, however, let the City go on with this practice while the plant underwent an environmental assessment. The charges were never picked up again.

In the early 1990s, a group of women who came to be known as the Sewage Sisters managed to infiltrate Toronto's sewage-treatment bureaucracy, with the help of the Province's newly

5 This charge was made in an August 5, 1993, letter from W. Gregson of the Ontario Ministry of the Environment to the Municipality of Metropolitan Toronto. Gregson was a provincial director under Section 53 of the Ontario Water Resources Act. Many provincial staff members were fired by the Harris-led Tories in the mid-1990s.

4 *Chlorine tank car, Ashbridges Bay Wastewater Treatment Plant, 1995. In the half-decade period before Peter MacCallum took this photograph, prominent ecologists, the IJC and the American Public Health Association all spoke out against the release of chlorine to the environment in any quantity. In 1999 the City of Toronto made a commitment to phasing out use of this persistent toxic chemical at the Ashbridges Bay Wastewater Treatment Plant. Yet as late as 2008, chlorine-filled railway tank cars were still rolling through the plant's gates.*

5 *Ashbridges Bay Wastewater Treatment Plant, 1959. Canada's largest sewage treatment plant got its start here, on landfill in Ashbridge's Bay, in the 1910s. By 2001 the plant was processing the wastes of some 1.2 million Torontonians, with a rated secondary sewage treatment capacity of 818 million litres per day. During 'rainwater events,' however, sewage-laden waters were (and continue to be) bypassed directly to Lake Ontario at the plant's seawall gates, its late-twentieth-century history being one of 'excessive toxicity' due to indiscriminate chlorine use. Such 'end-of-pipe' treatment practices became the subject of rising environmental concern in the early 1990s.*

crafted Environmental Assessment Act [6].[6] The municipal government of the day knew that if it did not recommend a bump up of the Ashbridges Bay environmental assessment to a level that gave citizens full standing, the provincial government would act to force the issue. Backed by prominent ecologists and inspired by environmental activism across the Great Lakes, the Sewage Sisters led a successful citizen-based effort to have the sewage-sludge incinerator at Ashbridges Bay shut down and convinced Council to adopt a cutting-edge sewer-use bylaw in 2000. One of the key objectives of the bylaw was to reduce the private dumping of toxic chemicals into Toronto's wastewater streams. But the continued dumping of toxic chlorine by the City itself amounted to civic hypocrisy.

As of the mid-1990s, the Sewage Sisters also began pushing for the substitution of chlorine with ultraviolet light treatment

6 This term was coined by *NOW* writer and activist Wayne Roberts. Still at work on various fronts, the Sewage Sisters initially comprised Karey Shinn of the Public Committee for Safe Sewage Treatment in Metropolitan Toronto and Debra Kyles of the Coalition for a Green Economic Recovery. By 1993, one of their key collaborators was Karen Buck of Citizens for a Safe Environment, along with many other citizen associates of various genders.

6 *The Sewage Sisters, Debra Kyles and Karey Shinn, were played by Jennifer Gauthier and Annette Paiement in Clay and Paper Theatre's musical production of* The Ballad of Garrison Creek, *written by Larry Lewis and performed during the summer of 1998 in various Toronto parks. Left to right: Jennifer Gauthier, Debra Kyles, Annette Paiement, Karey Shinn.*

systems at Ashbridges Bay. By the late 1990s, a promised plant-scale pilot study on the use of ultraviolet irradiation for disinfection was deemed an engineering success story. And then, virtually nothing. We are still waiting for viable alternatives to the use of tonnes of chlorine (approximately 1,980 kilograms per day at Ashbridges Bay, according to a 1997 draft environmental assessment). Fifty-tonne chlorine tanker cars continue to regularly supply the plant, nearly a decade after the April 1999 mediator's report on the Ashbridges Bay Wastewater Treatment Plant environmental assessment, whose seventh resolution set out the City's commitment 'to implementing an ultraviolet disinfection system for the . . . Treatment Plant effluent and discontinuing the use of chlorine disinfection.'

Toronto's claims of being an environmental leader, and the slowness of actions around a now-decades-old struggle to put wastewater treatment on a better environmental track, are clearly mismatched. The Province finally approved the Ashbridges Bay environmental assessment in 2008, seventeen years after it was initiated. Follow-through on the assessment's chlorine-reduction and phase-out provisions will be contingent on our taking more responsibility for the water that circulates in and through us. In an urban environment, our bodies are that part of nature that we directly inhabit – aided and abetted by water-filtration plants, sewage-treatment plants and the pipes and reservoirs in between. Unfortunately, the inclusion of citizens in infrastructure decision-making has given way to 'exclusionary moves' by City officials, according to Sewage Sister

Karey Shinn. Attempts in 2008 to stack the plant's Approval, Implementation, Compliance and Monitoring Committee 'against citizens' is not the way to go, Shinn states, 'especially given the ecological challenges that remain to be addressed in and around Canada's largest sewage treatment plant.'

Stronger engagement between urban infrastructure and the political ecology of urban nature is needed if we are to democratically sustain the environments that sustain us. Some 40 million people currently live in the Great Lakes–St. Lawrence Basin, largely in urban settings. If Toronto is to become a truly Great Lakes city, we need an urban political ecology that will make life better here and in basin areas downstream of our city – that is, both within and outside of the water supply and sewage-treatment systems discussed above. The Rosehill residents who fought the capping of their reservoir wanted more than such systems could offer them in the mid-twentieth century. With the recent intersection of precautionary theory, understanding and practice, perhaps more will be possible in the early twenty-first century. The stakes for urban health and the urban environment are certainly more important.

Books such as Devra Davis's *Secret History of the War on Cancer* continue to build the case that working with the symptoms of cancer can take us only so far. Now is the time, given the precautionary principle and our expanded knowledge base, to confront the environmental and ecological roots of this scourge. I say 'confront' because our systems entail interrelated private interests and public inertias. Davis goes so far as to reveal that some of the powerful corporations that have helped fund the war on cancer are implicated in its causes as chemical manufacturers. And we need to go further too: activism at the scale of the Great Lakes Basin needs to be complemented by more focused local forms of urban political ecology. Meaningful starts have been made on this front. The ball is now in the court of our current City Council, possibly to be passed back to us, pending action on previous commitments to detoxify Toronto's wastewater flows.

RECOMMENDED READING

Lee Botts and Paul Muldoon, *Evolution of the Great Lakes Water Quality Agreement* (East Lansing: Michigan State University Press, 2005)

Devra Davis, *Secret History of the War on Cancer* (New York: Basic Books, 2007)

Stephen Garrod, *Mediator's Report: City of Toronto Environmental Assessment, Main Treatment Plant*, 1999. Available at the Toronto Reference Library.

*Safe Sewage News*. (Toronto: Public Committee for Safe Sewage Treatment). This will hopefully soon be found at an archive near you.

Sandra Steingraber, *Living Downstream: A Scientist's Personal Investigation of Cancer and the Environment* (New York: Vintage Books, 1998)

Joe Thorton, *Pandora's Poison: Chlorine, Health, and a New Environmental Strategy* (Cambridge, MA: MIT Press, 2000)

Kerry Whiteside, *Precautionary Politics, Principle and Practice in Confronting Environmental Risk* (Cambridge, MA: MIT Press, 2006)

Wayne Reeves

# Addition and subtraction: The brook, the ravine and the waterworks

On a snowy afternoon in December 1875, engineer John Dent admired the coal smoke billowing from his tall chimney at the base of the Davenport Ridge, near the corner of Poplar Plains and Davenport roads. The machinery in his brand-new waterworks was running smoothly, pumping water through newly laid pipes to the residents and businesses of the Village of Yorkville. Along the bluff edge above him sat the mansions of the Toronto elite, which bore such fanciful names as North Holme, Rathnelly and Oaklands [1]. Dent wondered what the well-heeled residents of those estates thought of his smoky handiwork and his industrial intrusion into their pastoral landscape.

Two people looking down at Dent might have had mixed feelings about his project. Neither was supplied with water from the new operation, though both had a stake in the waterworks' early success. The Hon. William McMaster, owner of Rathnelly, had sold part of his land to the Village for the project. Samuel Nordheimer, at North Holme, was even more involved, having allowed Yorkville to draw water from his pond – the dammed-up Castle Frank Brook – for distribution by the waterworks. Perhaps the men felt some regret about the blighting of their views, or perhaps they took the smoke as an emblem of local progress.

The intertwined fates of Castle Frank Brook, Nordheimer Ravine and the Yorkville waterworks perfectly illustrate the complex interplay of nature and culture in the Toronto landscape over the last 135 years. Addition and subtraction are key themes. While a water-supply element (Dent's pumphouse) grew in size and importance, the natural features that were its raison d'être (a stream and its ravine) diminished over time, with change in both cases driven by population growth and demands for new, bigger infrastructure to support that growth. Local connections between nature and culture were close in the nineteenth century and all but sundered in the twentieth. Efforts have, however, been made in recent years to 'add back' the brook and the ravine to the landscape, pointing to a new relationship between nature, culture and water.

When Dent started his project, Castle Frank Brook rose in three small creeks in the Lawrence and Dufferin area and flowed

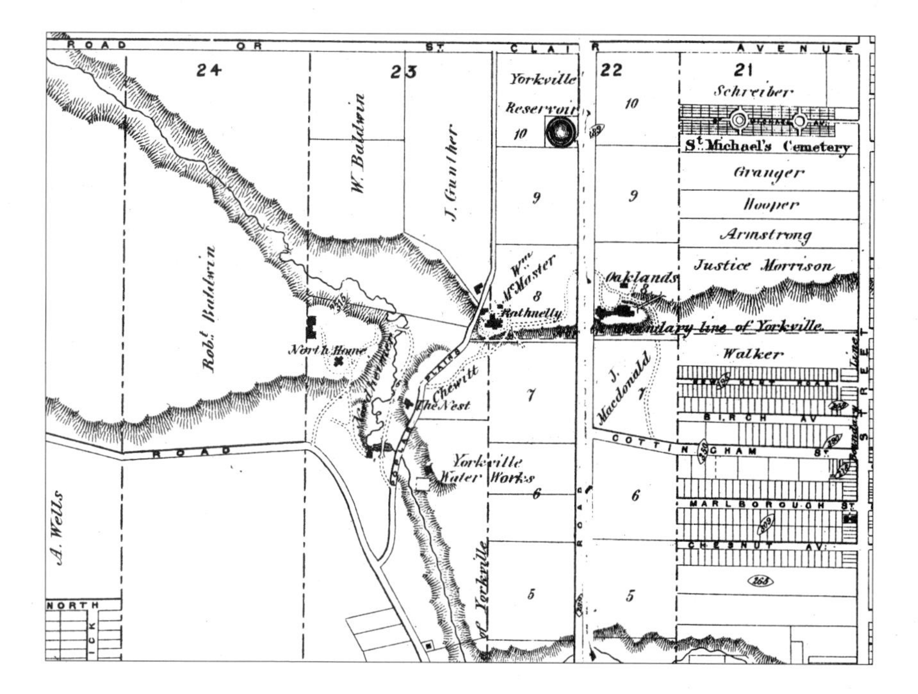


**1** *The Yorkville Water Works (now High Level Pumping Station) in relation to Castle Frank Brook and Nordheimer Ravine, 1878. The Nordheimer mansion is misspelled 'North Home.' The ravine and the brook cut through the Davenport Ridge (the Lake Iroquois bluff), which is represented by hachures along an east/west axis.*

southeast about 12 kilometres to the Don River south of where the Prince Edward Viaduct now is. The brook was named after Castle Frank, built in 1794 as the summer home and country retreat of Lieutenant-Governor John Graves Simcoe and his family. Situated on a high ridge that Simcoe's wife, Elizabeth, hoped would 'secure us from Musquetos,' Castle Frank afforded a view of both the brook and the river, which met less than half a kilometre from the house at what is now Bayview Avenue and Rosedale Valley Road. Though little surface water can be found today along the stream's course, its erosive force is manifest in three surviving topographical features. The most northerly is Cedarvale Ravine, the most southerly is Rosedale Valley and between the two lies Nordheimer Ravine.

Samuel Nordheimer settled with his brother, Abraham, in Toronto in 1844, after stints in New York and Kingston. Together, they set up A. & S. Nordheimer Co., selling sheet music, importing pianos and, according to the *Encyclopedia of Music in Canada*, becoming 'Canada's pioneer music specialty house.' Business boomed: the company began manufacturing pianos about 1890 and had built 11,000 units by 1910. By that date, Samuel was heading the renamed Nordheimer Piano & Music Co. and had achieved prominence in other fields, serving as the president of the Federal Bank of Canada, a director of Confederation Life, Toronto General Trusts and the Canada Permanent Loan and the German consul for Ontario. He was also president of the Toronto Philharmonic Society, organizer of the Chamber Music Association, chairman of the Toronto section of the Cycle of Music Festivals and promoter of major concerts in the city. Impressive achievements for a cultural outsider – a Bavarian Jew.

Unlike Abraham, who co-founded Toronto's first Jewish cemetery in 1849 and developed strong ties to local Jews, Samuel assimilated into the city's predominantly Christian establishment. He chose not to identify himself as Jewish in the 1861 census and associated little with Toronto's Jewish community. In 1871, he joined the Christian elite when he married Edith Boulton at St. James' Cathedral and became a member of the Church of England.[1] By 1872, Nordheimer's ascending social position found a landscape correlate. He and Edith moved into North Holme, a new thirty-five-room Second Empire mansion on the 25.5-acre estate he named 'Glenedyth' to honour his wife and the forested setting.

Glenedyth was strategically sited high on the Iroquois bluff, following the tradition set by other Toronto estates like Russell Hill, Spadina and Davenport. (Casa Loma became the benchmark for this approach shortly after Nordheimer's death.) As at Castle Frank, the land around Glenedyth fell away dramatically from the mansion on two sides, giving both long and short views. South of the sugarloaf lot was the low-lying Iroquois plain in which the city nestled; the vista also took in Lake Ontario and the plume of Niagara Falls. The close-up view to the north was dominated by the ravine that took Nordheimer's name. With the ravine landscaped to picturesque ends, Glenedyth embodied the image of an English country estate. Nordheimer achieved this in part by damming Castle Frank Brook to create a large pond and waterfall [2]. Editor James Timperlake gushed over the result in his 1877 book, *Illustrated Toronto*: 'The approach to the mansion is so constructed that as one crosses the many rustic bridges with their silvery stream bubbling underneath, it makes one fancy he is in fairy land.'

Though the Nordheimer Ravine is cited in planning documents by the late 1940s, it's not clear when the toponym came into common use. 'There is . . . no historical or geographical justification for the use of this term,' exclaimed a testy Austin

1 Boulton traced her lineage back to the Family Compact – the wealthy, conservative, Anglican elite of Upper Canada – and its prominent Town of York member D'Arcy Boulton Jr., who built the Grange in 1817.

Seton Thompson in his 1975 history, *Spadina: A Story of Old Toronto*. Perhaps he was piqued that the ravine wasn't named after Russell Hill (the 1818 house Nordheimer demolished to make way for his own) or the Baldwin family (original owners of Russell Hill). In fact, Glenedyth stood far longer than Russell Hill, and the Nordheimer family resided there longer than the Baldwins. Nordheimer's property straddled the ravine's east entrance, blocking the only easy access to it from Poplar Plains Road. As a nod to Toronto's complex cultural politics and a rare early recognition of a religious minority, Nordheimer Ravine would seem to wear its name well.

2 *The dam on the Nordheimer estate west of Poplar Plains Road, with Castle Frank Brook in the foreground, c. 1907.*

■ *Yorkville Water Works, c. 1880. The combined engineer's house/pumping station has since been remodelled and is still used as a private residence, rented out by the City at market rates.*

In any event, it was Nordheimer and his pond that provided a source of water for the Village of Yorkville. Isolated from Lake Ontario and with no major public wells, Yorkville was the first and likely only community in Toronto to use a stream for its municipal water supply. In 1874, Village Council balked at the Toronto Water Works Commission's terms for buying water from the City. A year later, Yorkville embarked on the construction of its own system, spending $75,000 by 1877 to build a pumphouse, a reservoir, a firehall and water mains – which meant arranging deals with Nordheimer and McMaster and hiring Dent to operate part of the system. In an interesting twist on water governance, Council first had to purchase the Yorkville Water Works Company's charter for $540.74 – the total value of this failed enterprise's assets and expenditures since its formation in 1872.

The combined pumping station and engineer's house, completed in 1875, was ideally placed. It was close to what seemed to be a reliable water source and to village customers (despite being located in York Township, just west of the Yorkville boundary). A retaining basin was dug beside the pumphouse [■], to which Castle Frank Brook apparently delivered more than just water: in 1878, the Village bought fifteen live pike for $4 to gobble up smaller fish in the basin.

Initial satisfaction with the project soon waned, however, with more consequences for local governance. In 1876, the *Directory of the Village of Yorkville* praised Council's 'public entrepreneurship' in having obtained 'a constant supply of pure water' for domestic purposes and firefighting. 'The water being obtained from a living stream is remarkably pure,' commented the directory's editors. But, as Yorkville's population grew – more than doubling between 1871 and 1881 to reach 4,825 in the latter year – so did discontent with the supply.[1] The Toronto *Daily News* darkly reported that Village water 'had been a source of evil in the town,' suggesting disease. Economic progress, public health, politics and water soon became an incendiary mix, as historian C. Pelham Mulvany summed up in his 1884 book, *Toronto: Past and Present*:

> The Yorkville Waterworks . . . proved a total failure, both the quantity and quality of the water were condemned by public opinion . . . The badness of the water supply at last caused so much deterioration in the value of Yorkville property that the annexation movement, long resisted by the village municipal officers, gained strength and was carried by a unanimous vote in 1883. Yorkville is now St. Paul's Ward, and enjoys the advantages of city water supply . . .

After annexation, Yorkville's pumphouse was seamlessly integrated into the City's lake-based water network, forming a distribution triangle with the John Street Pumping Station and the Rosehill Reservoir. The site was renamed High Level Pumping Station by 1885, a reference both to its elevation and its function in pushing water up on top of the Iroquois bluff to other newly annexed territories.

Plant expansion was required to deliver on this promised water supply. In 1889, American engineers Rudolf Hering and Samuel M. Gray noted that 'many of the more elevated parts [of Toronto] cannot be supplied with water on account of the lack of pumping capacity, which causes much annoyance to the people inhabiting them.' The City responded quickly with a new pumphouse at High Level, remodelling the original facility into a house that the engineer no longer had to share with his pumps and boilers.

This 1889–90 project was the first of many expansions at High Level undertaken to keep pace with Toronto's growth. Additional property was purchased on six occasions between 1887 and 1931 to provide space for new construction. Today, two structures stand at High Level: the remodelled 1875 pumphouse, and a much larger building, shaped like an inverted F and unified by the use of red brick and sandstone trim, which is made up of a central block built in 1906 and

1 Discontent at the waterworks also had a personal dimension. Village Council minutes for 1878 record John Dent suing 'for damages alleged to have been sustained through his losing possession of dwelling at pumping house.'

[4] *Built confidently in the Edwardian Classical style, the central block extension at High Level Pumping Station, c. 1906. The second pumphouse (1889–90; now demolished) is on the right.*

additions built in 1910–11, 1914–15, 1923–24 and 1952–53 [4]. Only the second pumphouse (1889–90) and a 150-foot chimney (1914–15) have been removed entirely.

Aside from gaining an ornamental fountain and a protective fence, the retaining basin at High Level Pumping Station remained intact throughout the 1889–1953 building program [5]. However, its function changed after the City takeover. A plan from 1898 shows the basin still taking Castle Frank Brook water from Glenedyth, but only to provide cooling water to High Level's pumps. In return, the basin received discharges from the station's condensers, air pumps, boilers and mud screens. Overflow and main discharge pipes linked the basin to Castle Frank Brook, which ran down the west side of the station property and then disappeared into a sewer at MacPherson Avenue.

No further exterior additions were made to High Level after 1953, when the property became part of the metropolitan water-supply system. While this might suggest that High Level's role had diminished, the opposite was in fact the case. Besides continuing to pump water, High Level is now the 'transmission control centre,' directing the system-wide delivery of tap water to over 3 million people in Toronto and southern York Region. In this system, four water-treatment plants on the lakeshore process raw water; eighteen pumping stations send the pure product through 510 kilometres of trunk water mains and 5,523

kilometres of distribution water mains to ten underground reservoirs and four elevated storage tanks, and on to consumers.

[5] *High Level Pumping Station from south of the retaining basin, 1925. All of these elements (dating from between 1875 and 1924) survive, except for the basin and the chimney. Part of the engineer's house is visible on the far right. Not in this view is the Style Moderne addition from 1952–53, which can be seen from Poplar Plains Road.*

By the time High Level reached its point of maximum development, Castle Frank Brook upstream had been irrevocably changed. When Samuel and Edith Nordheimer died in 1912, Glenedyth was left to their son, Roy. After Roy's death in 1921, the Nordheimer estate registered a subdivision plan carving up Glenedyth [6]. The 1923 plan saluted the Nordheimers with Roycroft Road (now part of Roycroft Park), Boulton Drive and Glen Edyth Drive, which traces an original carriageway to North Holme. The less fortunate mansion was demolished in 1929.

The Glen Edyth subdivision had little impact on the deep ravine cut into the Iroquois bluff, though it sealed the fate of Castle Frank Brook to the south. Local water and sewer lines were extended into the area and covered by Boulton Road; the remainder of the meandering stream bed was levelled with fill.

Development of the low-lying section of the Glen Edyth subdivision was slow. By the beginning of 1927, only two houses had been built along Boulton north of Cottingham Road. The first resident on the block, William L. Caldow, likely never knew his house at No. 144 lay astride a buried stream. Despite the small size of the lots, the Depression and World War II stymied further construction. Most of Boulton's east side was not built on until 1945–50; the west side came even later, between 1950

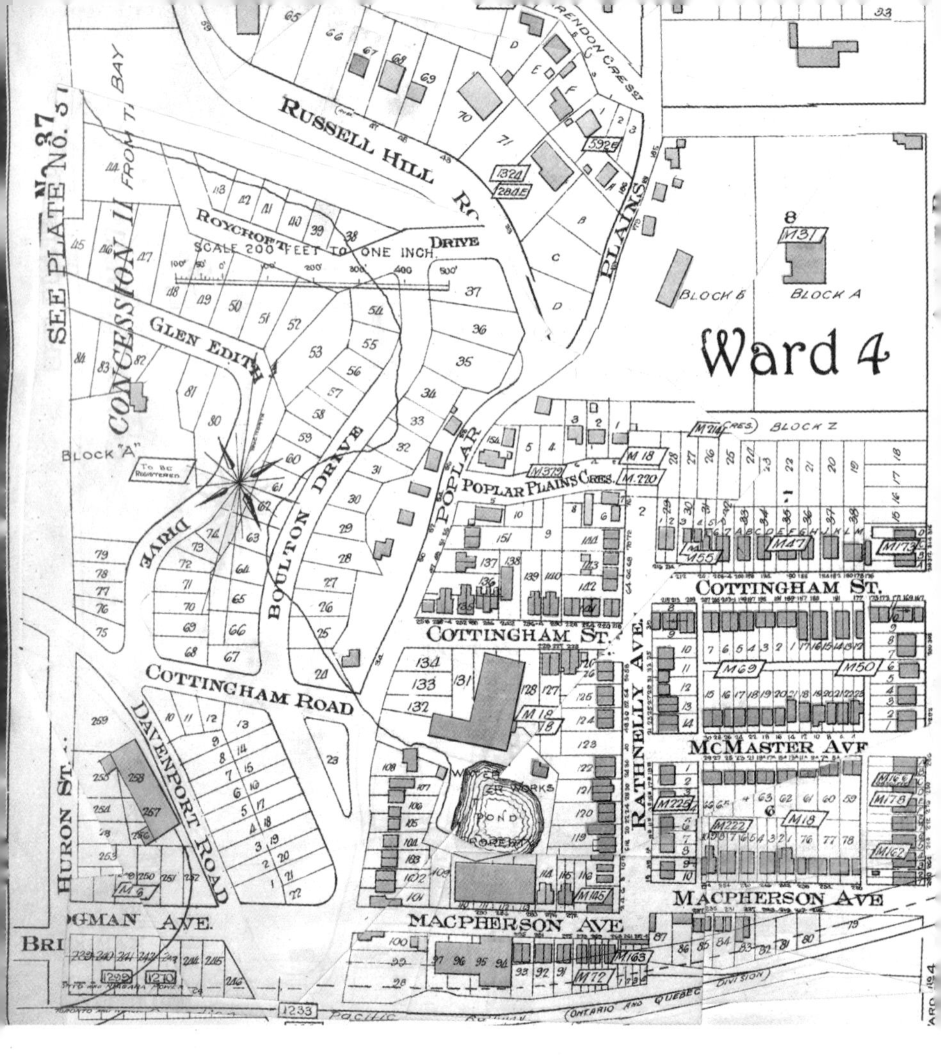


6 *Detail from Charles Goad's* Atlas of the City of Toronto and Suburbs *showing the new Glen Edyth subdivision on the old Nordheimer estate, 1923. Castle Frank Brook runs southeasterly out of the Nordheimer Ravine and down across Boulton Drive. The latter was built in 1926, pushing the creek into a combined sewer. High Level Pumping Station is at the lower centre.*

and '55, the modestly scaled houses contrasting with the toney surroundings. Land within the Glen Edyth subdivision was acquired by the City on a lot-by-lot basis between 1930 and 1953 to produce a chain of parkettes in the area. The Roycroft Road allowance, extending some 182 metres into the ravine from Boulton Drive, was finally closed by City bylaw in 1971 and dedicated as parkland in 1992.

The closure of Roycroft Road came too late to save either the brook or the ravine, which were impacted by big infrastructure projects. Castle Frank Brook was repeatedly dug up and reburied to lay new water mains between High Level and the St. Clair Reservoir – by 1941, Boulton Drive was underlaid by four water mains, ranging from fifteen to 137 centimetres in diameter – but the *coup de grâce* was delivered by drainage work required in advance of the construction of the proposed Spadina Expressway. Nordheimer Ravine lay on the roadway's route, which required a massive reorganization of stormwater in central Toronto. In 1965, engineers told City Council that the Spadina South Trunk Drain Sewer 'would be constructed by the cut-and-cover method and the existing open watercourse will be eliminated by the new drain. Upon completion of the drain, we propose to grade and grass the valley floor to restore the park-like atmosphere of the ravine.' This work was completed by 1970.

The threat posed by the Spadina Expressway came to an end in 1971, though more cut-and-cover work undertaken for its transportation successor, the Spadina subway, impacted both Nordheimer Ravine west of Spadina Road and all of Cedarvale Ravine. There, the grades were raised and the valley floor flattened, and any semblance of natural water flow was eliminated by 1978. This came despite pleas from community activists for the ravines' protection in *Foxes and Watercress: A Proposal for the Cedarvale-Nordheimer Ravine* (1972). A rather more gentle intervention came in 1988, when a gravel path was finally laid up Nordheimer Ravine to connect Boulton Drive with Sir Winston Churchill Park atop the St. Clair Reservoir.

Castle Frank Brook survived a little longer on the High Level grounds than it did within the Glen Edyth subdivision. The stream appears on a 1934 fire-insurance plan, but is gone by 1943. The pumping-station pond – the last aqueous tie to the Nordheimers – was not filled in until 1963–64. Whether it was reduced to a mere decorative feature in its waning years is unclear. The stream and the pond are now covered by a flat expanse of turf, topped by playground equipment first installed at the request of the Rathnelly Area Residents' Association in 1969.

Flowing water in Nordheimer Ravine has been reduced to a ditch dribble perched on fill. The ditch collects just the runoff of the north ravine slope rather than the brook's entire upper watershed; the water vanishes into a catchbasin before reaching Boulton Drive. Runoff on the south side of the ravine has found a higher ecological use in the Glen Edyth and Roycroft Wetlands, two small restoration projects of the Task Force to Bring Back the Don and the City of Toronto. Constructed in 1997–98 to form hardwood swamps, these sites have been planted with black ash, red maple, marsh marigold, skunk cabbage, Bebb's sedge, common rush and other plants that enjoy low, moist conditions. They are cared for by a community stewardship team. As long-time team leader Susan Aaron puts it, her volunteers 'assist the wild to become the everyday.'

The historic connections between the brook, the ravine and the waterworks can now be divined only through the printed word and archival images. As individual landscape elements, their current integrity is highly variable. High Level has come out the best, boasting the oldest and best-preserved sequence of buildings in Toronto's water-supply system, with architectural quality that harkens back to an era of noble public works. Nordheimer Ravine occupies the middle ground. Samuel Nordheimer's landscaping is entirely gone, dug up and filled in, time and time again, yet the ravine still surprises and delights on the walk up Boulton Drive. The road approaches and then parallels the Iroquois bluff; off to the left a deep cleft appears in the bluff's face, a passageway into postglacial erosion, Edith's glen. And then there is Castle Frank Brook, the epitome of subtraction. Even there, the disappointment of loss is tempered by the possibilities of at least limited restoration nearby on the ravine floor.

And what of the water-based interplay between nature and culture in the local landscape? The two reached peak equilibrium in the mid-1870s. Water-supply needs drove High Level to have a close functional tie with Castle Frank Brook and Nordheimer Ravine; the latter's amenity value made it a perfect site for the ambitious Nordheimer family. But community demands soon tipped the balance away from nature. Castle Frank Brook could not deliver the quality and quantity of water Yorkville needed, and once High Level began drawing and delivering lake water in the mid-1880s, the stream's existence became less and less essential. Beginning in the 1920s and lasting for nearly five decades, Nordheimer Ravine increasingly became a mere development opportunity for private housing and public infrastructure. The balance began shifting back toward nature in the early 1970s and then accelerated in the 1990s. The story of the brook, the ravine and the waterworks might appear to be unique and idiosyncratic, but it fits within a larger pattern of how Toronto has interacted with its water over time.

# Explorations

Nordheimer Ravine
375

Shawn Micallef

# Subterranean Toronto: Where the masquerading lakes lay

Though we complain that Toronto is cut off from Lake Ontario by expressways and railroads, we are, in fact, surrounded by water. Toronto is criss-crossed by ravines and streams just out of view of city streets and by buried creeks that could be underfoot at any time. Yet there is another feature that goes unnoticed: Toronto possesses ten inland subterranean lakes that masquerade as public parks on the surface. These lakes are not natural, wide-bodied versions of buried creeks, but rather human-made reservoirs that keep Toronto supplied with drinking water and water pressure. Unless you know where they are or can read subtle landscape hints like a civil engineer, they will pass under your radar, if not your feet. Once you know they're there, though, it's impossible to think of the land in the same way: the massive presence of what lies beneath is overwhelming.

Rosehill Reservoir and the Lawrence Reservoir and Pumping Station are two of these covered and secret lakes – engineered aquifers that represent two very different eras in the growth of Toronto. The former, built in the mid-Victorian era when Toronto was just beginning to expand beyond its early boundaries, is connected to various historical and topographical figures and features in central Toronto. The latter, built eighty-five years later, came at a time of rapid and massive

urban expansion, when Metro Toronto's post-war manifest-destiny project ate virgin farmland by the acre.

[1] *The newly completed-Rosehill Reservoir, 1875.*

Rosehill Reservoir [1] lies southeast of the busy intersection of Yonge and St. Clair, a very uptown kind of place surrounded by a quintessential Toronto mix of old single-family homes and modern high-rise apartment buildings. Rosehill Avenue leads east from Yonge along the north side of the reservoir until it reaches a ravine edge. Climb a few steps to the top of the reservoir and the entire city to the south is suddenly visible, demonstrating why this is an ideal location for a reservoir: it's on high ground. Located on the crest of ancient Lake Iroquois, it sits above much of downtown Toronto, simplifying water delivery. And though the hill is still there, the only roses around are the ones on the lapels of smartly dressed uptown men.

The high ground at Rosehill was chosen in 1872 to ensure efficient fire protection for the whole city. Not only did it provide enough water pressure to protect even the highest warehouse buildings, the City's chief engineer at the time even suggested the reservoir would render steam fire engines unnecessary. In addition, this new inland lake would be located next to the Vale of Avoca Ravine and its bubbling brook – a connection to 'natural' water flow that exists to this day.

**2** *The Rosehill Reservoir grounds, then known as Reservoir Park, c. 1910. A staircase leading to the top of the reservoir is visible in the distance on the right.*

Securing the reservoir site involved the purchase of two estates whose names live on in Toronto nomenclature. Jesse Ketchum, a wealthy tanner who posthumously lent his name to a Yorkville school, owned 200 acres south of St. Clair between Yonge Street and what is now Bayview Avenue. After his daughter Anne married banker Walter Rose, they were given the land and called it Rose Hill. After Rose died, the property was divided and, in 1865, Joseph Jackes (of Jackes Avenue fame) bought part of it. Jackes would sell half of his site to the City; the remainder was built up with solid single-family homes.

The Rosehill area lies just north of Summerhill, a neighbourhood named after an estate acquired by businessman and University of Toronto vice-chancellor Larratt Smith in 1867. In *Young Mr. Smith in Upper Canada*, his daughter, Mary Larratt Smith, describes a thirty-room mansion, behind which stood stables, a cow barn and a henhouse:

> At the back of the property were two orchards; beyond them, the woods ran down to the spring-fed creek in the ravine below... The old house was pulled down in 1909 to make a subdivision, Summer Hill Gardens. The street, Summerhill Avenue (once a part of the property), is still there, but the city reservoir in the park where my father and my uncles used to race beetles and wash their dogs when no one was looking has been covered with cement.

Smith's property had previously belonged to stagecoach- and steamboat-line owner Charles Thompson, who had created an early amusement ground on part of his land. In her book *Aristocratic Toronto: 19th Century Grandeur*, Lucy Booth Martyn writes that 'attractive paths led down to the clear running stream spanned by a narrow wooden footbridge at the bottom of the ravine.' Though Smith closed the amusement ground, public use of the ravine would return.

Smith sold a portion of his land to the City in 1872 for the reservoir, stipulating that part of the land must always be maintained as a public park. Smith's legacy is the narrow rectangle of parkland leading south from the reservoir into the Summer Hill Gardens neighbourhood, along with the ravine area that comprises David A. Balfour Park and extends east to Mount Pleasant Road.

Construction of the Rosehill Reservoir began in October 1873. For the first two-thirds of its life it was uncovered, truly an inland city lake, and was a place of recreation for Torontonians. Photos from the early 1900s show a well-tended walkway around the reservoir edge reminiscent of the Central Park Reservoir running track made famous in the 1970s film *Marathon Man*. However, instead of a sweaty and paranoid Dustin Hoffman, the Rosehill promenade was frequented by prim and proper Torontonians in formal attire. The amusement-park grounds became known as Reservoir Park, and included lush and wild spaces, much like today, as well as manicured gardens, a floral clock and open spaces where those good Torontonians, still in their Sunday best, played what appear to be spirited games of baseball [2].

After World War II, the area became less bucolic. Mount Pleasant Road was opened as a major artery that cut Rosedale in two. Apart from dramatically altering the topography of the Vale of Avoca and Reservoir Park, the new road meant the destruction of a number of stately Rosedale homes. Aerial photos show a nearly rectangular, jet-black 'lake' bulging to the southeast to follow the ravine edge, surrounded on the other three sides by houses. Yet, like much of Toronto, everything about this neighbourhood was about to change, rapidly and radically.

In 1956, the only high-rise at Yonge and St. Clair was just east of the new subway station, and the reservoir was still situated in the midst of a thick urban forest of mature and leafy trees.

3 *Rosehill during covering, c. 1966.*

Even as late as 1964, most of the residential neighbourhood was intact, but within a year a number of lots were bulldozed to make way for the towers that exist today.

Until this period, the reservoir had remained open to the air – allowing Larratt Smith to wash his Victorian dogs – but by the 1960s, several reasons compelled Metro to roof over its finest inland lake. A 1954 report by Gore & Storrie Consulting Engineers stated that 'the open Rosehill Reservoir should be covered and made into a two compartment reservoir, similar to the St. Clair Reservoir, with its capacity increased from 35 to 50 million gallons.'

However, capacity was not the only reason to cap this lake. The City was also looking to save money by not having to chlorinate the reservoir's water a second time. Maintenance could also be reduced; in 1942, civic officials reported that when Rosehill had last been cleaned, 143 truckloads 'of muck and weeds were taken away,' and several thousand fish 'went down the drain.' The Toronto Archives also holds letters sent by the Commissioner of Works in 1946 asking the fathers of four boys to 'have a talk' with their sons about the dangers of trespassing on reservoir property. The boys had been found inside the fence surrounding the reservoir, suggesting the urban infiltration movement was alive and well before mid-century.

Apart from these 'traditional' nuisances, fears associated with the modern age contributed to the capping. According to Michael McMahon's exhibit *Pipe Dreams: A History of Water and Sewer Infrastructure in Metropolitan Toronto*, 'nuclear fallout,

4 *Rosehill after covering, c. 1975.*

growing pollution from auto emissions, and pollution from plants, humans, and other animal species, provided added rationales for the covering of the reservoir. Some of these fears need to be seen in context, particularly in the wake of the Cuban missile crisis of 1961.'

Yet when the City covered the reservoir, it went beyond what might be expected of conservative Toronto the Good: the plan included a landscaped park complete with children's play facilities, reflecting ponds, a fountain and even a 'water cascade.' The covering project began in 1964; by August of 1967, it was nearly complete, and sod was being laid. At the same time, the neighbourhood was changing. Many of the remaining houses on the north and west sides were demolished and replaced with high-rises. In the distance, across Rosedale, cranes were building another of Toronto's great modern projects: St. James Town.

Today Rosehill is much like it was in 1967. A central fountain feeds a shallow, lazy river that flows into two large blue ponds before crashing over a series of small waterfalls on the north side of the park. The fancy and coiffed dogs of Summerhill run and jump together on the grass – always a little soggy, the poor drainage betraying the concrete lid lying only a few inches below – while their owners stand chatting in little clusters, oblivious to the body of water underfoot. The park's centrepiece is a fountain with an atomic sculpture (a water molecule?) held up by stainless-steel arms, still shining forty years on [5]. Visitors can sit on the curved cement bench that looks like an

**5** *Rosehill's centennial sculpture fountain, 2007.*

oversized and petrified living room chesterfield and contemplate the atomic age. The '1967' and a Canadian centennial symbol inlaid in the raised patio are intact, but a small podium with empty screw holes and a ghost outline is missing its plaque.

Perhaps it's better that the plaque is gone because it could not compete with the plaques one could imagine existed here: 'On this spot, Pierre Elliot Trudeau declared that Canada came of age; that peace, order and good government may long reign; that the state has no business in the bedrooms of the nation,' and so on. Rosehill's design so evokes the optimism of the super-sexy Trudeaumania era, it seems as if it were designed as a rest stop on the way to Expo '67. The skyline has changed a bit since then, but the commanding view of the city below, the lush ravine on the east and the constant hum of Mount Pleasant beyond remain as they were in the sixties.

Before the modern age radically transformed this landscape, Avoca Avenue once ran down into the ravine to what is now east Rosedale. Today, a steep footpath follows the route down to Yellow Creek. Directly below the reservoir, Swiss Family Robinson–style paths and staircases also descend to the ravine floor. Like that artificial lake high above, the ravine here is a deceptively natural landscape. Carved by the creek over the last 10,000 years, the Vale of Avoca has also been greatly altered by humans. The stream has been channelized, the construction of Mount Pleasant Road has shifted the ravine walls and manhole covers have appeared along the dirt paths, a sign of infrastructure below.

On summer nights, when the only light that trickles in is from houses perched on the ravine rim, this area is well-travelled cruising ground for men looking for sex. It's been like this for decades: there are old stories of Toronto police officers descending into the darkness with flashbulbs that they would fire off into the dilated pupils of the men who were in for an unpleasant evening in one of Toronto the Good's moral slammers. Today, a handful of folks loiter on the western edge of the reservoir most nights, waiting for the right person to come along and accompany them into the dark ravine. Should you wander down old Avoca Road you might think you're alone, but the red embers from burning cigarettes and the rustle of branches a few feet away indicate somebody is near. They, along with the dog walkers, joggers, Ultimate Frisbee players and moms pushing strollers, keep the Rosehill landscape in use day and night, while below the grass and concrete our trusted water infrastructure, working away behind the scenes, keeps Toronto running.

Across town, at Lawrence and Caledonia, another secret Toronto lake plays a similar role for the city. Unlike Rosehill, overlooked by middle-class and wealthy apartment and condo dwellers, the Lawrence Reservoir and Pumping Station serves an unsung working-class neighbourhood. Surprisingly, and also unlike Rosehill, Lawrence can boast that it's a listed heritage property, despite being less than fifty years old.

Lawrence Reservoir was among Metro Toronto's first major projects, constructed in 1958-59 and conceived as a covered facility from the beginning [6 7]. The same report that recommended covering Rosehill also proposed building a new reservoir on Lawrence Avenue West between Keele and Dufferin to slake current thirsts and anticipate further growth. Though it was built in the heroic post-war era, Lawrence lacks Rosehill's atomic-age accoutrements. Its vast and plain lawn is marked only by a few small stands of trees that provide shade for the fans of teams playing baseball on the adjacent diamonds or cricket atop the reservoir. The all-white clothing worn by the cricket players is a formal echo of those dressed-up Victorian folks who enjoyed sport at Rosehill [8]. The woodlot beside the diamonds is a tiny landscape analogue to the Vale of Avoca. No water flows here – there's just a collection of swampy puddles after the rain – but it likely provides space for the untold adventures of area children.

In lieu of shiny towers and a downtown skyline panorama, Lawrence commands views of low-rise industrial land, nondescript residential areas and discount home-furnishing and automotive shops. To the northeast, across the retail wasteland that is Orfus Road, Yorkdale Mall is visible, the original Eaton's and Simpsons concrete boxes intact, the suburban landmarks that are possible only with a steady flow of pure water. Though Lawrence's setting is not as dramatic as Rosehill's, it too is on high ground. The city slopes gently southward from this point, but the reservoir is also located on the civic equivalent of the continental divide – the lofty meeting point of Toronto's Don and Humber watersheds.

Lawrence lacks Rosehill's long, rich history and slowly evolving landscape. The Lawrence site was still typical Ontario

6 *Site and surroundings of Lawrence Reservoir and Pumping Station, c. 1960. The pumping station, with its two stone owls, is at the lower left; the valve house is at the lower centre.*

farmland in the 1950s, its homesteaders missing from the memoirs of the city's early aristocracy. Archival photos from the fifties show new light-industrial buildings along Caledonia, still just a dirt track in places, messy and muddy from construction vehicles. Companies like Kelyie Mason Floor Covering and Stirling Page Industries Ltd. occupied new low-rise buildings, big cars parked in gravel lots outside. Typical of Toronto's outer suburbs, the area around Lawrence was part of the country's rapidly expanding economic engine, going from farmland to fully built-out form in just a few years. North York expanded so quickly that public wells could not keep up; there are stories of mothers bathing their children in ginger ale due to the lack of potable water. Though Lawrence anticipated further growth in Metro Toronto by providing infrastructure before industry and housing arrived, its lake water also served some very immediate needs.

Lawrence's nod to the magnificent public works of previous generations (and the source of its heritage designation) are two limestone buildings designed in the early modern style of the day. Both have subtle features that harken back to the art deco era. These solid-looking yet elegant buildings played an important symbolic role in the idea of progress: public works like the Lawrence Pumping Station gave the impression that suburban growth was being built on a solid foundation, countering the appearance of flimsy private development. Trimmed in black labradorite, the main entrance to the larger building is rather grand, but a 'No Soliciting' sign and tightly pulled blinds suggest visitors are not welcome. Most curious are two stone owls perched atop obelisks that stick out of large concrete dishes. Though owls seem like odd additions to the front of a waterworks building, having a pair of such creatures standing guard may not be such a bad idea. Owls are associated with wisdom, knowledge and protection – all things we want from a waterworks – and the Egyptian hieroglyphic representation of water was three owl symbols in a row. Apart from any deeper meaning, it is simply nice to find these sculptural forms in a part of town that lacks both whimsy and beauty. We can thank Toronto's post-war engineers for breaking out of their narrowly technical world and giving us such thoughtful pause.

It's worth walking the perimeter of these two reservoirs to get some grasp of the scale of these works and the hidden volume they hold. Our reservoirs are so well-integrated into the fabric of the city that it's easy to forget their important primary role and instead think of them simply as parks. Perhaps that's as it should be, but since water is such a fundamental part of life, there may be a danger in doing so. Canada has been rocked by several drinking-water scares and tragedies in recent years. Shouldn't we always have some reminder in the city of the massive undertaking necessary to supply us with millions of litres of the clean stuff every day? Perhaps those ever-vigilant owls at Lawrence, and the molecule at Rosehill, are reminder enough of the valuable resource we keep buried underground in Toronto.

7 TOP *Lawrence Pumping Station, main entrance, c. 1960.*

8 ABOVE *Cricket practice at Caledonia North, with the Lawrence Reservoir embankment and valve house in the background, 2008.*

David A. Robertson & Andrew M. Stewart

# The Garrison Creek mouth and the Queen's Wharf: Digging up 200 years of shoreline development

Toronto's archaeological resources are constantly threatened by the ongoing success of the city. By necessity, cities continually rebuild upon themselves as part of their cycle of regeneration and growth. They are the most complex of archaeological sites because they are the products of prolonged and dynamic patterns of human activity, always in a state of formation.

Toronto's waterfront has been a significant site of settlement, transportation and industrial activity. Over the last two centuries, development and other landscape interventions have transformed the shoreline at the entrance to Toronto Harbour, and the surviving heritage fabric tells many important stories about the city's history. There is no resemblance between today's Lake Ontario shoreline and the one that existed when John Graves Simcoe, Lieutenant-Governor of Upper Canada, established the military garrison of York at the mouth of Garrison Creek in 1793. But this change did not happen all at once. Rather, the shore has gone through a series of changes, of which the condominium boom and the public-space revitalization plans are just the latest.

Archaeology is one means by which this historical sequence of development along the waterfront can be investigated. In fact, it is now often required as part of the development process. Some such archaeological work has been carried out near the now-buried mouth of Garrison Creek and in the once-offshore area at the foot of Bathurst Street occupied by the Queen's Wharf in the nineteenth century [1]. Although more efforts will be required as redevelopment plans within the area progress, the results to date provide tangible material evidence of this legacy of change.

The shore first emerged from the receding waters of glacial Lake Iroquois 12,500 years ago as a flatland of deep-water-laid clays and shallow-water-laid sands. This plain extended south from the bluff that runs parallel to Davenport Road well out into what is now the bed of Lake Ontario. It was dissected by rivers and dotted with giant boulders carried along by the glaciers and then abandoned during their retreat. Over the following millennia, lake levels fluctuated, and the landward portion of the plain was

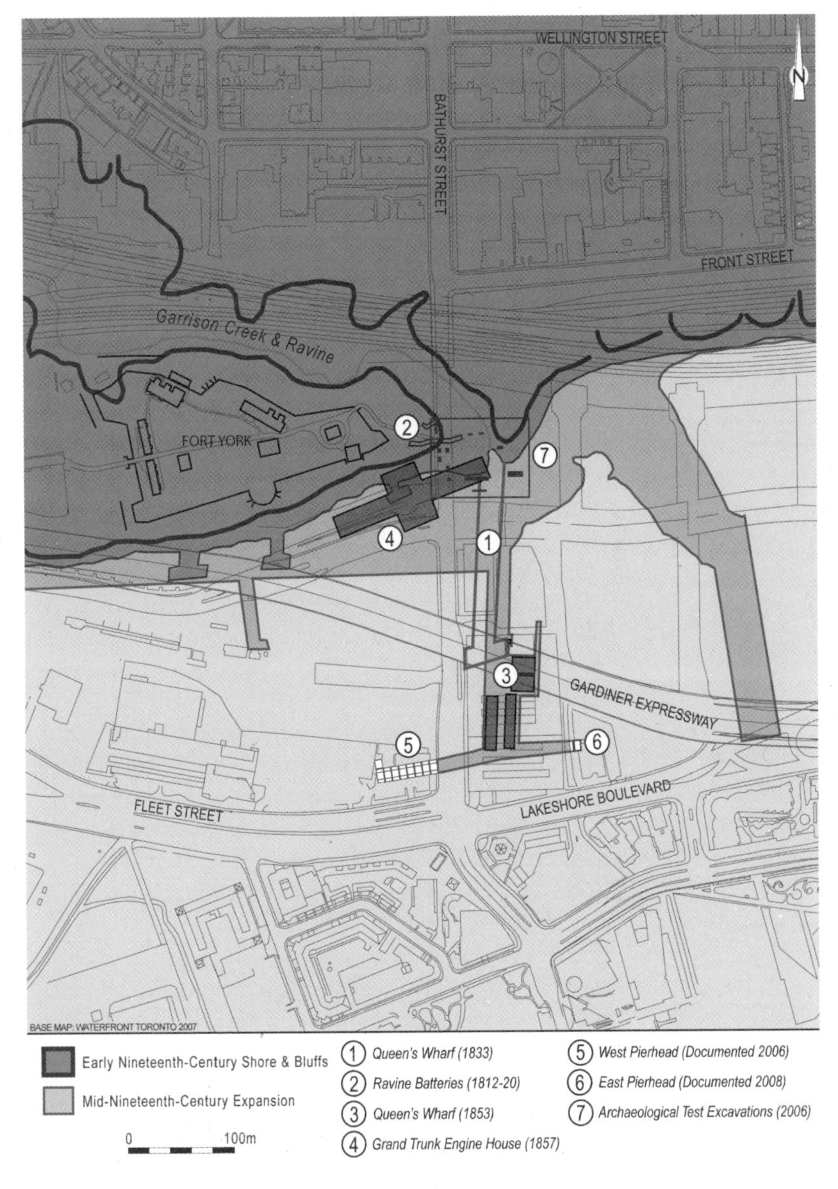


*1 Selected features in the historic development at the mouth of Garrison Creek and areas of archaeological investigation.*

colonized by a succession of forest communities, so that by the time of the founding of Fort York, the lakeshore was occupied by a dense forest of white pine and balsam fir, interspersed with a variety of hardwoods. By 1793, the shore itself was a narrow limestone shingle beach below a steep shore cliff that was some six to eight metres high. Extensive wetlands existed at the mouths of the various creeks that flowed into the bay.

The site Toronto would be founded on attracted Simcoe's attention because of the potential it offered as a protected harbour, a key consideration given that he believed war with the United States was both inevitable and imminent. At that time, the future harbour was an almost completely enclosed bay, protected by a low sandy spit (now Toronto Island). The only entrance was through a narrow passage opposite the mouth of a stream that meandered through a wide, deep ravine and emptied into the lake near the current intersection of Bathurst and Front streets. The defence of the harbour, and the town that sprang up on the shore just west of the Don River, depended on the further fortification of the western entrance to the bay. The lands at the mouth of the stream, which was eventually called Garrison Creek for its military associations, were suitable for just this purpose.

The first British fortifications, established on a site cleared by 100 Queen's Rangers (today's Fort York), were set on a defensible promontory of land, with the lake to the south and the ravine, through which Garrison Creek meandered, to the north and east. The initial work involved widening the mouth of the creek to accommodate bateaux and a wharf. As the western flank of the fort was more vulnerable, it required strengthening through the construction of earthworks, palisades or stockades, artillery batteries on the mainland and a blockhouse at Gibraltar Point, which was on the end of the spit, on the far side of the harbour entrance.[1] Though the government was reluctant to commit the necessary funds to the construction of the western defences, over the next few years, the military complex grew to span the tablelands on both sides of the creek and even the floor of the ravine, both at the mouth and further upstream on the north side of Fort York. However, many of these upstream installations were swept away by a spring flood in 1852.

Some of the ravine features at the mouth of the creek, and an artillery battery and its associated earthworks, were the target of small-scale archaeological excavations carried out in 2006, when ten test trenches were dug on the lands to the east of the Bathurst Street bridge and under the bridge. There was hope that these investigations would provide information

1 Confusingly, the name Gibraltar Point has migrated over the years to the south end of Toronto Island. In the 1790s, Gibraltar Point was roughly where the Hanlan's Point clothing-optional beach is today.

on the original character of the creek mouth and lakeshore zone, and that they could be used to chart some of the later historical developments in the area, which had resulted in far more drastic alterations to the landscape of the creek mouth. The first of these changes was the construction of the Queen's Wharf, a pier that stretched over 200 metres into the lake. Originally known as the 'New Pier,' the Queen's Wharf was first constructed around 1833, on the east side of Garrison Creek. The wharf had a dual purpose: in addition to serving as a docking and cargo-handling facility, it was intended to hinder the growth of the offshore sandbar that continually threatened to block off entry to the harbour by speeding up the flow of water through that entrance. In 1837, the wharf was lengthened to almost 250 metres so as to reach deeper waters, and a seventy-five-metre east pier head was added at the same time. These changes allowed the wharf to service larger vessels. The wharf was widened again in the early 1850s, and a 120-metre west pier head was added as well. By the late 1880s, the wharf had grown to cover an area of over nine hectares.

2 *The depth of the lakefill and its unstable qualities meant that only limited glimpses of the timbers making up the Queen's Wharf were observed during the test excavations. Here, two sets of beams can be seen running across the trench below the scales.*

The preliminary archaeological work to investigate the Queen's Wharf was limited in scale relative to the features it was intended to investigate, and was further hampered by a number of factors stemming from the fact that the excavations took place, literally, in the former lake. The remains of the wharf were buried by over 2.5 metres of fill and were submerged in water. Digging to these depths required the use of heavy machinery such as backhoes, but the fills were so wet and loosely packed that the walls of the trenches frequently collapsed within seconds of being excavated. The archaeologists could only quickly glimpse the features lying at the surface of the groundwater and interpret them as best they could before they were reburied by slumping soil [2].

Despite these difficulties, the investigations were successful in highlighting the potential survival of elements of the former shoreline and creek bed. In one trench, the flood plain of Garrison Creek may have been found, while another trench yielded tantalizing evidence, in the form of a timber platform and some early-nineteenth-century brick fragments [3], that may mark the location of the ravine artillery battery, though further excavations will be necessary to date the structure or to identify it with confidence. Yet another trench revealed riverine or lakeshore gravels. These gravel layers contained

▪3 *Early-nineteenth-century handmade bricks recovered during the test excavations. The brick on the left may be from an artillery battery built in the ravine after the American attack on York in April 1813. The brick on the right, which has been smoothed by water action, may be from the fortifications that were built on the east side of the creek in the 1790s, but which were largely abandoned after the Battle of York.*

large quantities of artifacts, many of which were water-worn, either from the current of the creek or shoreline wave action. Household ceramic artifacts found included fragments of dishes and plates of the types and styles used from the end of the eighteenth century to the mid-nineteenth century. There were large numbers of leather shoe fragments [▪4], two of which had squared toes – a style fashionable in the mid-nineteenth century – as well as pieces of scrap leather. A relatively complete handmade brick was found, very similar in size and style to those used in the construction of the early-nineteenth-century buildings at Fort York, and of other roughly contemporary archaeological sites in Toronto such as the first parliament buildings at Parliament and Front streets and the original Toronto General Hospital at King and John streets. Food waste included fragments of large animal bones, some of which had been butchered. Finally, there were large quantities of wood debris, much of which was clearly waste from cutting and/or squaring timbers using axes and adzes.

Many nineteenth-century accounts describe the creek and its ravine as an open sewer that was clogged with waste prior to the filling of the upstream portions of the ravine and the construction of the Garrison Creek sewer in the 1880s. So the artifacts recovered from this trench may come from various sources. The ceramics and food waste could be related to either the military occupations at Fort York or to residents further up the creek. The preponderance of shoe remains, including obvious manufacturing waste, suggests that this material is likely debris from a civilian cobbler's operation, presumably located upstream. The woodworking debris may be related to work at the fort, construction of the early phase of the Queen's Wharf or perhaps other events further upstream.

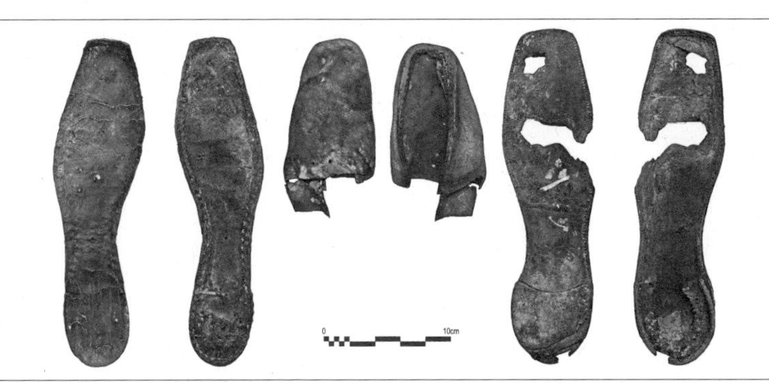


[4] *Examples of shoes recovered from the gravel layer during the test excavations. It is surprising to find so many examples from so limited excavations. Perhaps they are discards from a cobbler's shop located upstream.*

Several other trenches uncovered large timbers that are clearly related to the Queen's Wharf, although little pattern could be seen in terms of the wharf's overall construction. Fortunately, recent discoveries on some adjacent properties provide a far better indication of the form of the wharf and the methods used in its construction than did the targeted archaeological excavations. Portions of both the east and west pier heads were uncovered during separate construction projects on sites on either side of Bathurst at Fleet Street between 2006 and 2008, and were fully documented by archaeologists once they were exposed [5]. The remains found on these construction sites were extremely well-preserved, hinting at what will be found during further excavations at the mouth of Garrison Creek.

Although the Queen's Wharf marked a new era in the use of the Garrison Creek area, the most sweeping change in the lakeshore was brought about by an ambitious campaign of landmaking on the part of the railway companies and City authorities who, in the 1850s, began to extend the waterfront into the lake, burying what little remained of its bluffs, creek mouths and nearshore environment to create a featureless plain for industry, railroads, piers and slips. The new water's edge was defined by a wall of timber cribs similar to those used in the Queen's Wharf, behind which were dumped vast quantities of sewage, 'cellar dirt' excavated from town construction sites and, most importantly, soil cut from the south edge of the original shoreline bluffs, including the promontory on the east side of the creek on which many of the Fort York buildings had stood. To the south of the now landlocked remains of Fort York, on the west side of the creek, the Grand Trunk Railway constructed a cruciform-shaped engine house with a turntable, freight house,

5 *The west pier head of the Queen's Wharf was exposed on a construction site at the northwest corner of Bathurst and Fleet streets in 2006. This section of the wharf dates to the 1850s and 1860s. These crib structures lay below the waterline, acting as foundations for the deck of the pier.*

smithy, temporary shed, pumping house, carriage house and shed, wharf and temporary passenger station.

The test excavations also revealed aspects of this period of harbour development. Most obvious were the massive layers of sand, silt and clay fill that were used to raise the surface of the newly created lands two metres above the level of the lake. But the work also uncovered a probable footing from the 1850s engine house and a section of one of its walls, indicating that substantial remains of the foundations may have survived the demolition of the building and later developments within the area.

The separation of Toronto from its waterfront is documented in these vestiges of features found at the mouth of Garrison Creek, as well as in the remains of other structures along Toronto's nineteenth-century industrialized waterfront. They are the origin of the considerable barrier that grew between the people of Toronto and the lake on which their city was founded. This development took place despite the fact that, from the early 1800s, numerous plans called for the preservation and enhancement of the waterfront as public recreational space and that much of the lakeshore had been laid aside for such public good. While the growth of this industrial district was the source

of considerable prosperity for the city, it sparked controversy as well. Period sources frequently comment on the railways having left filling jobs incomplete, resulting in the formation of putrid lagoons or unfilled sloughs that trapped raw sewage from the city's drains and trash thrown into creeks. They also complained that open spaces on the waterfront were used to store huge stockpiles of coal, leading to wind-borne clouds of coal dust spreading a pall throughout the harbour and city, and that the railway lines that ran along the waterfront all but cut off pedestrian access to the shore. The complaints raised 150 years ago are little different from those of our time.

The waterfront was, from the outset, a contested landscape. The construction of the Gardiner Expressway and the abandonment or under-utilization of the waterfront industrial and railway lands during the mid-twentieth century, as well as the intensive commercial and residential redevelopment projects of the past twenty years, have all led to similar conflicts. The latest round of change, however, has provided archaeologists with opportunities to examine some of the contentious developments from earlier ages and to help ensure that this history is remembered.

Michael Harrison

# The vanishing creeks of South Etobicoke

While most streams in central Toronto were degraded and buried in sewers during the 1800s, many of the original creeks of south Etobicoke survived well into the mid-twentieth century. Located a considerable distance from the new capital of York, these creeks survived the deforestation of the area by pioneers clearing land for crops in the early to mid-nineteenth century, as well as the urban growth and development that characterized the late nineteenth and early twentieth centuries. Though diminished and degraded, these creeks – North, Jackson, Superior and Bonar – continued to flow through the landscape, their ravines remaining a distinct feature of the area well into the twentieth century, when they finally succumbed to urbanization and the perception – real and imagined – that they were polluted and unsafe. However, glimpses of many of these creeks are still visible today if you know where to look. And we soon may be able to see more than just glimpses, if residents and politicians are able to harness new opportunities to partially reintegrate these creeks into the landscape by daylighting portions long buried beneath the urban fabric.

The 1811 patent map that lists the original landowners in what is now south Etobicoke provides an excellent view of the creeks that existed at that time, including the aforementioned North, Jackson, Superior and Bonar creeks [1]. The lives of these watercourses are inextricably linked to the use of the land within their watersheds. Originally covered by thick forests, these watersheds evolved over thousands of years to become a finely tuned and balanced system that produced a steady flow of cool, clean water abounding in sensitive cold-water fish species such as salmon. But it was the forests that first attracted Lieutenant-Governor Simcoe to the area.

Though far from the new capital, the land along the Humber River contained excellent stands of timber, which the new town required for the construction of homes and public buildings. One of Simcoe's first actions was to establish a sawmill on the west bank of the Humber River in 1793 to take advantage of the superior wood found in Etobicoke township, which he

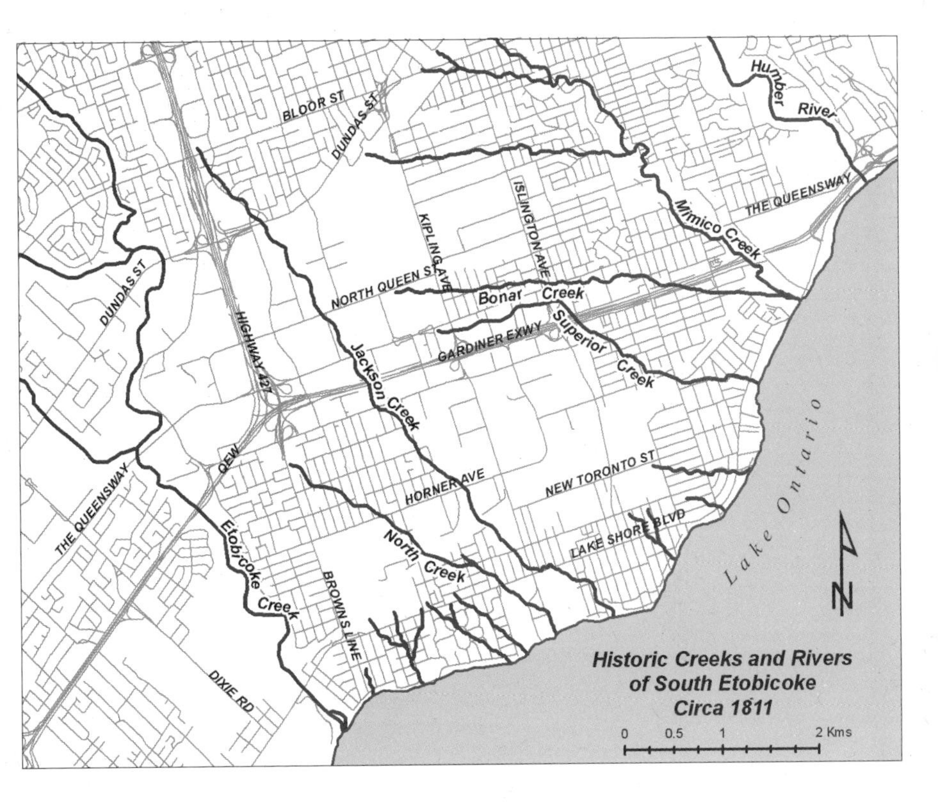


1 *South Etobicoke, c. 1811.*

protected for the Crown by creating the King's Mill Reserve. In the ensuing years, roads and bridges were built; the Lake Shore Road, which was laid out in 1798 over an Aboriginal trail that ran through south Etobicoke, became the main road connection between York and Niagara; adjacent roads were hacked out of the forest along the concession lines to serve the newly surveyed lots to the north; and bridges were built over many watercourses.

In his 1799 book, *A Short Topographical Description of His Majesty's Province of Upper Canada in North America*, Sir David Smyth, surveyor general, provides a good description of the area shortly after settlement began:

> A little to the westward of the garrison are the remains of the old French fort Toronto; adjoining to which is a deep bay, that receives the river Humber, on which are saw mills belonging to government . . . Further to the westward (that is, between the Humber and the head of Lake Ontario) the Tobycocke [Etobicoke], the Credit, and two other rivers,

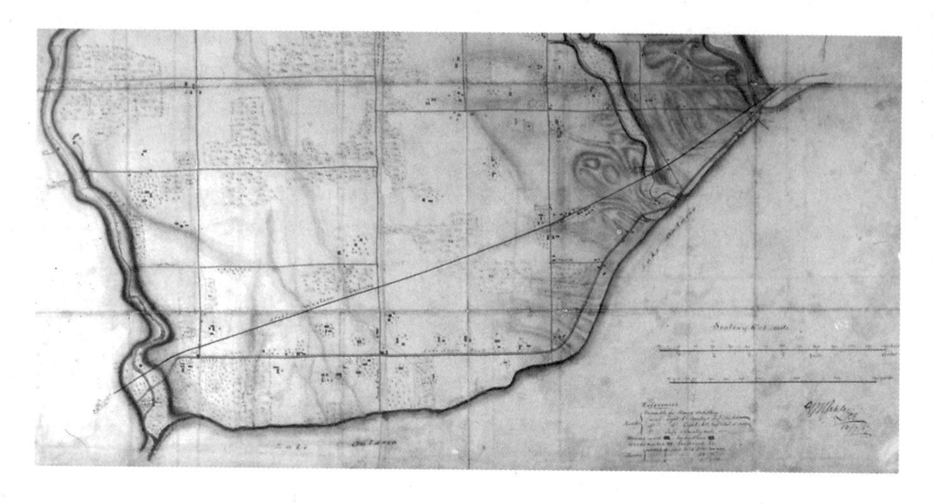

2 *'Toronto District Sketch Sheets of a Reconnaissance in the Country between the Rivers Humber and Etobicoke from the shore of Lake Ontario to Dundas Street on the north,' 1867.*

> with a great many smaller streams, join the main waters of the lake; they all abound in fish, particularly in salmon . . .

As York was built up, the need for wood became voracious. Deforestation in the King's Mill Reserve accelerated, and the government continued to enlarge the reserve to try to ensure a constant supply. Beginning as 726 acres in 1793, the reserve was enlarged to 850 acres in 1799 and then to 1,181 acres in 1803. By 1822, most of the trees had been cut down. As the forests were cleared, the finely tuned system of steady cold-water flows in the creeks began to break down.

By about 1840, the Township of Etobicoke had generally completed the pioneer stage of its development and had emerged into an era of settled agriculture. In the early 1840s, forest cover dipped below 50 percent, and the rivers and streams began to flood during storms as natural systems broke down. The first severe floods in the Humber River and Etobicoke and Mimico creeks were recorded in 1850. Other floods followed in 1851, 1859, 1878 and 1880, each more severe than the last, damaging mills, washing away dams, destroying property and resulting in loss of life. In her 1974 book *Etobicoke: From Furrow to Borough*, Esther Heyes wrote that in 1878, there were twenty-six gristmills and thirty-seven sawmills on the Humber, and that 'many of these were damaged beyond repair in the flood.' Though undocumented, it is expected that similar impacts would have been felt in the North, Jackson, Superior and Bonar creeks.

In 1855, the first railway, the Great Western, was built through south Etobicoke on its way to Hamilton, resulting in the first subdivision plan being registered in the area. Bridges, culverts and viaducts were constructed to facilitate this major engineering project [2]. As a result, the path of Bonar Creek was altered, and the valley was partially filled to provide for a bridge crossing over Mimico Creek. Not until the 1990s would it be discovered that the fill was contaminated with heavy metals.

Most of the land in south Etobicoke had been cleared by 1867, and over the years the remaining woodlots steadily decreased, with the exception of the 500 acres of pine and oak owned by the Smith family at the mouth of Etobicoke Creek. In 1871, however, Samuel Smith Jr. sold the land to James Eastwood, who found ready buyers for the valuable timber. All of the trees were cut down within a few years, and by the mid-1880s, the remaining forest cover was limited to the major valleys of the Etobicoke and Mimico creeks and the Humber River.

Between 1881 and 1891, the population of Toronto more than doubled in size, jumping from 86,000 to 181,000. Land speculators bought up agricultural land and registered plans for subdivisions as new residents moved into the surrounding areas. It was at this time that south Etobicoke began its transformation from an agricultural community to an urban/industrial suburb. Long Branch Park, with over 200 cottage lots, was built on seventy-five acres of the old Smith Estate in 1884, and was quickly followed by the provincially funded Victoria Industrial School in 1887 and the Mimico Asylum in 1888 (later renamed the Lakeshore Psychiatric Hospital, and now known as the Hospital Grounds). North and Jackson creeks flowed through the Hospital Grounds [3], and both watercourses and their ravines were repeatedly modified in the early twentieth century to facilitate agricultural uses west of Kipling Avenue and ornamental landscaping to the east.[1]

In 1890, the industrial suburb of New Toronto was founded on 550 acres of land near present-day Islington Avenue and Lake Shore Boulevard West. A radial streetcar line was built to serve the area along the Lake Shore Road in 1892, reaching the mouth of Etobicoke Creek by 1895 and providing quick access to the city. Speculators and developers ran amok, and by the 1910s most of south Etobicoke had been subdivided into building lots. Infrastructure improvements promoted more and more development: in 1917, the Lake Shore Road was

1 The Hospital closed in 1979. Today, most of its buildings are occupied by Humber College's Lakeshore Campus, surrounded by Colonel Samuel Smith Park.

3 *The Jackson Creek bridge in the Hospital Grounds, c. 1910.*

paved between Toronto and Hamilton and became Ontario's first highway. In 1928, it was widened, and the streetcar line was double tracked. At the same time, population increases allowed local communities to achieve political independence from Etobicoke Township. Mimico achieved village status in 1905, followed by New Toronto in 1913 and Long Branch in 1931. This new pattern of growth accelerated the deterioration of the local environment.

Prior to the end of World War II, residential and industrial development in the area had generally taken place within the existing landscape, and few wide-scale changes had been made. However, the post–World War II construction boom required more land for development. As a result, land originally seen as unsuitable was filled in and built upon or used for recreational purposes. Creeks that were in the way were forced into sewers, and their valleys were filled in. Lax pollution laws, too much development and limited capacity at old and antiquated sewage plants resulted in the discharge of sewage and industrial pollution into these watercourses. The result was public complaints and the inevitable push to have these offensive creeks placed in sewers and out of sight.

And so North Creek [4], which originally drained a large area of south Etobicoke, flowing in a southeasterly direction to Lake Ontario from its headwaters near the interchange of the

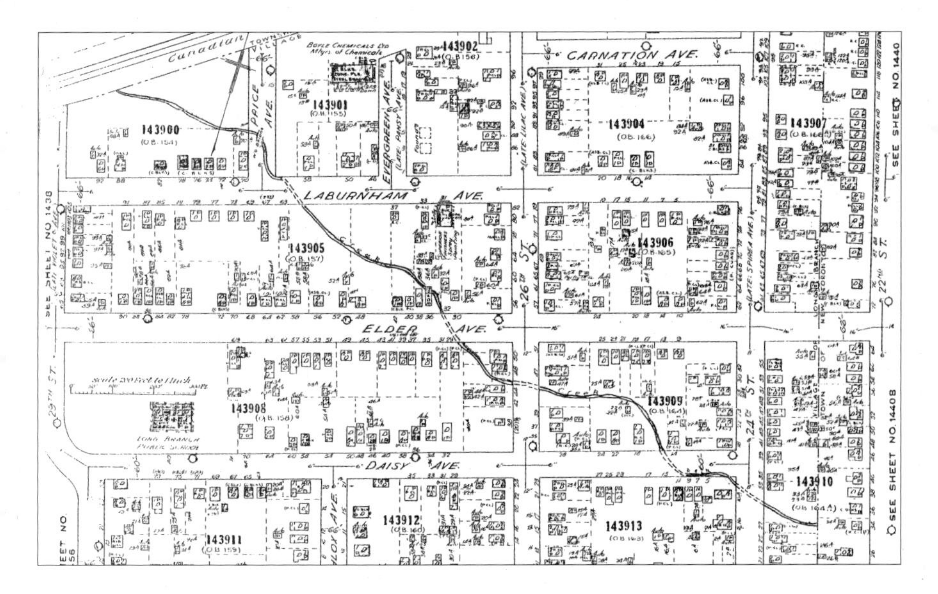


■ *North Creek: route through Long Branch, 1952.*

present-day Gardiner Expressway and Highway 427, became a remnant of its former self. Most of the creek was placed in a sewer sometime after 1958, following four years of complaints, as chronicled in the *Advertiser*, by residents that industry was dumping waste in the creek. At a Long Branch Council meeting in 1954, councillor Maurice Breen openly admitted that this was a problem, but explained that the local sewage plant was already overloaded. The Village 'could not allow the industries to send their huge quantities of water through the plant so at present they have no alternative but to deposit some liquid waste in the creek.'[2] After repeated calls for action from local council, the Etobicoke-Mimico Conservation Authority commissioned a report that resulted in the creek being placed in a sewer through Long Branch. Today it disappears into a sewer just south of Laburnham Park [5], popping up on the Hospital Grounds at the foot of Kipling Avenue. There, the creek is separated into two distinct portions: the northerly portion just south of Lake Shore Boulevard West flows intermittently through a man-made channel completed around 1930, while the southerly part of the creek – the most natural portion of the watercourse – flows through its original ravine, retaining much of its tree cover as it crosses Colonel Samuel Smith Park to enter Lake Ontario. Even though it has now been hijacked as a stormwater system for the Hospital Grounds, its flow is relatively constant, providing an important habitat for migrant

[5] ABOVE LEFT *North Creek in Laburnham Park, 2008.*

[6] ABOVE RIGHT *Jackson Creek near the Queensway and Atomic Avenue, 2008.*

birds like warblers, resident birds like black-crowned night herons and other wildlife including foxes and beavers.

Though highly modified, Jackson Creek, which also once drained a large area of south Etobicoke, originating just north of Bloor Street West near Highway 427 and flowing in a southeasterly direction to Lake Ontario, still flows through part of its upper reaches, surrounded by the industrial and commercial enterprises that moved in after World War II. Had this area been developed as a residential community, the creek would most likely have been buried. Instead, it was shunted off to the sides of large industrial and commercial properties, and though it has been channelized, it is now largely vegetated by trees and other riparian vegetation like bulrushes [6]. Further south, in the residential community of Alderwood, its course meanders through Douglas Park south of Evans Avenue, and along the western boundary of the now closed Alderwood Collegiate on Valermo Drive [7]. These two portions of the watercourse, totally devoid of any riparian vegetation, are essentially grassed ditches with intermittent flow. From there the creek was piped south through the Hospital Grounds [8] and through the residential community of New Toronto. East of the Hospital Grounds its ravine was filled in and houses were built over it. The ravine west of Eleventh Street was filled in and became a park, while the mouth of Jackson Creek to the east remained relatively undisturbed until 1947, when it was redirected into a sewer that empties into Lake Ontario at the foot of Eleventh Street. Its former ravine and rocky beach remained part of Rotary Park until 1958, when it was deemed unsafe by New Toronto Council and was covered with fill as part of a $90,000 park-improvement plan.

The lower reaches of Superior Creek, which originated near present-day Kipling Avenue, just north of the Gardiner Expressway, and flowed in a southeasterly direction to Lake

2 *The Advertiser*, June 3, 1954.

Ontario, were modified in the 1910s, but the upper portions of the creek continued to flow freely across north Mimico until construction of new homes after World War II led to citizen complaints over pollution and safety. Articles first began to appear in the *Advertiser* in the early 1950s, when the residents of north Mimico began to complain loudly to Mimico Council of perennial flooding. In December 1954, the Etobicoke-Mimico Conservation Authority commissioned a report on the feasibility of placing the creek in a sewer through Mimico, but the work does not appear to have been completed until sometime after 1965. Today, nothing remains of the creek, and cars now speed where it once flowed in a wide and steep ravine down Stanley Avenue south of Station Road.

7 ABOVE LEFT *Jackson Creek at Alderwood Collegiate, 2008.*

8 ABOVE RIGHT *Jackson Creek on the Hospital Grounds, 2008. The parapet footings are the only visible trace of the original bridge (see* 3*).*

Bonar Creek, a tributary of Mimico Creek, had its headwaters just north of Superior Creek and flowed in a southeasterly direction until it joined up with the main watercourse at its vast wetland at Lake Ontario. It continued to flow until about 1950, when most of it was placed in a sewer and parts of its former watercourse were filled in and topped with warehouses and a sewage-treatment plant. In 1957, the operators of the McGuiness Distillery, which was located east of Grand Avenue – now the Mystic Pointe development – began to fill in the remainder of the creek's ravine in order to construct a number of warehouses north of Manitoba Street. Today the creek is mostly channelized and only the lower portion of the creek, below the CNR rail line, flows above ground.

Many communities lament the loss of their historic creeks, and in the last decade a new urban movement seeking to daylight these watercourses and restore them to as natural a condition as possible has taken root. Examples of successful daylighting projects can be found in Victoria, Vancouver, Los Angeles,

Seattle, Yonkers, Liverpool, Zurich and Seoul. In 2000, the Colorado-based Rocky Mountain Institute published a study of the benefits of daylighting buried streams titled *Daylighting: New Life for Buried Streams*. The study reviews the challenges and costs of restoring buried streams and includes case studies of projects from around the world. The authors of the report state that in each case, these projects have enhanced public spaces, improved water quality, allowed for greater hydraulic capacity to alleviate flooding and provided for the restoration of natural habitats.

The lost creeks of south Etobicoke do not have to remain lost. In 1997, as part of a report titled *Toward the Ecological Restoration of South Etobicoke*, I proposed the restoration of the lower portion of Bonar Creek where it flows into Mimico Creek, and the large wetland that once existed there. The proposal attracted the attention of the Etobicoke-Mimico Watersheds Coalition, a group of volunteers supported by the TRCA that began planning for restoration in 1999. In 2003 the coalition proposed that the restoration of the lower portion of Bonar Creek become a project under the Wet Weather Flow Management Master Plan, to be developed by the TRCA and the City of Toronto. The Bonar Creek Wetland Project proposes to recreate the historic Mimico Creek wetland, east of Legion Road and north of Lake Shore Boulevard West [9]. The class environmental assessment is expected to be completed by the end of 2008, and implementation should begin in 2009.

There are other opportunities too. Jackson Creek has the most potential for daylighting, as its upper portions are still largely above ground. The TRCA and the City of Toronto should undertake a thorough analysis of this recently rediscovered creek and look at opportunities for restoration and enhancement, including daylighting the parts of the creek that flow under the Hospital Grounds and Rotary Park. Likewise, North Creek could be restored through Laburnham Park, and the buried portion of the creek that runs through the Hospital Grounds could be daylighted to enhance the park and provide additional habitat for wildlife. Restoring Superior Creek will be a challenge, as much of its former course is now totally built over, but one definite opportunity exists at the foot of Superior Avenue, where it could be daylighted where it once flowed into Lake Ontario.

[9] *The forested confluence of Bonar and Mimico creeks near Lake Ontario, 1889.*

These and other opportunities could be made possible by the Etobicoke Waterfront Stormwater Management Facilities Study Municipal Class Environmental Assessment that is currently being undertaken by the City of Toronto. The assessment examines the possibility of consolidating the thirty outfalls that currently discharge stormwater directly to Lake Ontario, and to improve water quality by implementing stormwater-management facilities. The restoration of these creeks and the wetlands that used to exist at their historic mouths would enhance water quality, help deal with flooding, aid in the cleaning of Toronto's waterfront, enhance public spaces and provide wildlife habitat.

There is hope that these creeks will once more flow across the landscape. The engineering expertise that buried them so recently could now be used to daylight them, providing both inspiration and lessons for the other lost creeks of Toronto. What once was could be again.

Murray
Seymour

# Streamscape: Rivers of life in the city

The ravines are secret places. There's no plot to keep them that way – they're just beneath our radar. More accurately, they're beneath our line of sight. But at the bottom of each there's water, and water is the basic requirement for life. If Mars has water, it's well-hidden, and we've found no life there. And then there's the moon: great potential, but no water – no life. No matter how cleverly we build, invent and organize, the fact is that the farther away from water we place ourselves, the greater our separation from life.

There are six major watersheds across the City of Toronto. Each has its own character; each is unique. From west to east, the land changes from the barely covered grey shale that underlies the whole region to a thicker overburden best seen in the Scarborough Bluffs. The west is within the Carolinian forest region, while the east is in the Great Lakes–St. Lawrence. Deciduous, broadleaf trees slowly give way to more evergreens such as pine, white cedar and balsam fir as you move across the city.

My wife and I have been exploring Toronto's ravines for thirty years. One warm spring day we were turning over slabs of shale in Etobicoke Creek, wetting them to better reveal possible fossils. The stream was a mere trickle that day, totally unlike how it was at the beginning of the nineteenth century, when a treaty between the Mississauga Indians and the British reserved salmon fishing rights for the natives. A salmon would have to have legs or wheels to get up the creek today.

The creek was also different that day than it was on the night back in August 1857 when stormwaters surged down the channel, sweeping up everything before them. The storm had already brought devastation to parts of Brampton upstream, but here, fortunately, there were fewer people and structures to sweep away. At least twenty people were not so fortunate almost a century later in 1947, when another flood of stormwater came roaring through and carried them off. It happened again in 1954 when Hurricane Hazel struck. Since then, the creek has been more controlled and channelled, but less extensively than its neighbour to the east.

*Highland Creek.*

Mimico Creek has become so urbanized that it is virtually invisible. Much of it is firmly ensconced between concrete banks, yet if you look closely, you can still see it rippling and gurgling over its bed of shale. The outstanding memory my wife and I have of this stream is looking down the sloping side of a cement retaining wall covered in a thin film of ice and shuddering at the feeling of total helplessness we would experience if by chance we happened to slip down into the rushing torrent below.

Then there's the Humber. Or, more accurately, the Humbers. The West Humber flows quietly from a dam that regulates its fluctuations, thereby escaping the ravages of water-level extremes. Instead of the crack willows that populate the flood plains of the other streams, here there are flowers and established vegetation. It is a much gentler river.

One day, descending into the valley of the West Humber from a suburban road, my wife and I jumped to the sound of a loud snort and the pounding of the hooves of a large deer bounding through the grass and red osier dogwood bushes. Earlier that day we had spotted a cottontail rabbit, a grouse and a great blue heron farther upstream. Ahead of us, beaver dams were waiting to be discovered. Toronto's streams attract a wide variety of creatures, and they are one of the most attractive characteristics of the ravines. They live there because of the plentiful food, the minimal human harassment and, of course, the presence of water.

The shale is still close to the surface in this part of the West Humber. We once spent an hour eating lunch beside a long, broad cascade of small shale steps where a little unnamed stream trickled and tinkled down in glistening sheets with diamond flashes. A place well worth searching for.

The main Humber has a variety of attractions too, not the least of which is the annual salmon run. Many times, my wife and I have stood captivated by the sight of a line of men in full waders standing in the river just downstream from the first weir above the Old Mill Bridge, casting their lines in hope of snagging a salmon. Every once in a while a line would go taut, a rod would bend and slowly, slowly, the point where the line entered the water would begin to move. The drama of fish and fisherman might continue for up to half an hour. Usually, it ended when the line went slack. Most fish were released to try again to leap over the just-too-tall dam. Occasionally one was coaxed to shallow water where it was netted and milked of its roe, which would be used as bait to catch other fish. The fisherman would then take it home to share all the mercury and other pollutants it had collected in its body with his happy family.

One day, we watched a very young couple emerging from a rocky little tributary, the girl carrying a salmon that had become caught between the boulders. Her face was shining with the gratitude of someone who has just received a gift of almost heavenly benevolence. Mercury or no, that little family would enjoy salmon for some time to come.

That's the summer Humber. But there's a winter Humber too. One freezing day we watched a flock of mallards quacking and burbling over a scattering of corn along the bank. More ducks came gliding in, but they hadn't yet adjusted to the idea of ice. Instead of plowing to a stop in the water as usual, they slid and slid and slid. They slid into other ducks, knocking them aside. There was much tumbling and quacking and rising up and down and fluttering of wings. Soon, however, everyone concentrated on the corn again.

But winter has a dark side. On a January exploration, we found a patch in the middle of the river where the water ran faster and ice had not yet formed. Beside it, a great blue heron stood at the edge watching, silent and immobile as only herons can be. A few equally silent snowflakes drifted down and dusted its back and wings. It might've been a vain hope that a fish would come near this small spot that kept him there, but it was the only hope the heron had.

A different great blue heron, this time on a log beside a pond close to the upper Don. It stood still and intent, then suddenly tilted its head so one eye could look down into the dark waters. With a great splash it dove. A moment later it emerged with spray flying and much flapping of wings. In its beak was a bright red fish. It tossed and caught the tidbit until it was properly positioned for swallowing.

Another day, we were feeding bread to mallards who were raising their families in the pond. Ducks are not the most fastidious eaters, and many crumbs escaped their beaks. Looking closer, we noticed darting shapes beneath each bird. The shapes were red. There are no red fish native to this part of the world. However, there are goldfish that, given the proper conditions, will become red. Why someone would release a pair or more of goldfish into this pond is anybody's guess. Yet these relatives of the carp are adaptable to a wide variety of habitats, and in this pond, a good-sized school of the creatures was apparently thriving. Between the birds and the fish, not a crumb was wasted.

*Humber River.*

Taylor Massey Creek is the main tributary of the Don within the city. There's a stretch of the creek in Warden Woods where the

ever-present hum of traffic is only barely audible, and it's possible to forget for a moment how much of this stream has been channelled, poisoned and clogged with debris. Here, it is green and wild. Wildflowers nod in the warm breezes, and the smell of grass, pollen and flowers floats enticingly. The constantly changing flow of glints and sparkles from the ripples of the moving water dazzle and enthrall. The tiny rippling sounds mingle with the buzz of insects, the trees' whispering leaves and the occasional bird calls. It's a Debussy afternoon, and about as close to heaven as may be found on earth.

But water is not always like this. Highland Creek and its tributaries rise and flow entirely within the city. They drain a smaller area than any of the other rivers, but when aroused, their power is every bit as impressive. One pleasant evening, the sky turned suddenly dark. The clouds roiled and churned, taking on a greenish cast. And then it rained. It was a tropical rain the likes of which we rarely see in these temperate zones. By the next day the creek we had waded through earlier had become a thundering torrent, and boulders the size of small cars were rolling majestically downstream. Here, we witnessed the power of the natural world unleashed, and saw and heard vivid proof that our efforts to control it are puny indeed. This river is younger than the others, and its course clearly shows that it is still scouring down to find its long-term level.

The Rouge streams are both the farthest east and the wildest of Toronto's rivers. Close to their banks, we've watched a herd of perhaps twenty deer browsing on new wild leeks. We remember where the lady's slippers grow in spring. We recall finding laid out in the snow the tale of an attack by coyotes on a young racoon. One of our most cherished memories is of the day after the once-in-fifty-years storm that scoured Highland Creek. We were sitting with a group of friends at the end of a high railroad embankment, looking down at the Rouge's frothing torrent. Slowly, we became aware of another sound – a small burbling, chortling sound. Up the hill waddled a string of Canada geese goslings, who had been attracted by our murmuring voices. Separated from their parents far upstream in the storm, they had somehow survived the boiling waters and managed to get up on dry land. They waddled right into the middle of our group, sat down, wriggled into comfortable positions and promptly

*Rouge River.*

fell asleep, exhausted by their struggle to stay alive. Some would say we should have left them there to let nature take its course, where they would probably have become dinner for a fox or coyote. Instead, we decided we too were a part of nature, so we picked them up and carried them to the zoo nearby.

What have we learned from our thirty years of exploring these ravines, besides the names of some plants, insects and animals and a few of their habits? My wife and I have seen that all things change. Rocks, plants, animals, weather, rivers are constantly becoming something else. Humans have accelerated this change. Thirty years ago, only a few tree huggers were concerned about our streams and ravines and the life they foster; now there are many who care deeply and are working hard to make them the vibrant heart of the city once more. It is our fervent hope that our grandchildren and their children will also have the opportunity to make memories of watching the goldfish swimming beneath the ducks and seeing deer grazing, of hearing the constant murmur of miniature waterfalls and smelling the sweet scents of summer flowers and fall apples. We hope that they too will have the chance to find themselves in the heart of Toronto.

Liz Forsberg & Georgia Ydreos

## Participation/precipitation: Can community-based arts help keep us afloat?

A river of people dressed in blue meanders through Toronto's streets and alleys in a performance of a long-lost creek. Wading pools across the city transform into spaces for participatory art installations, enticing park-goers to romp in a hollow of water balloons, stamp out wool to create felt or build aqueous sound sculptures. Youth draw elaborate sound maps of their acoustic environment on a public path alongside a stream, inviting an accidental audience to take more notice of their surroundings. A production on an ancient shoreline by a cast of 120 community members features a chorus of children singing, 'Let's start again, let's build it over. Set free the waters buried deep below. Set free the waters flowing through our dreams.'

In her book *Lure of the Local: Senses of Place in a Multicentered Society*, art historian Lucy Lippard asserts that artists have a unique capacity to foreground the breadth of their environment and society's relationships to it: 'Artists can make connections visible. They can guide us through sensuous, kinaesthetic responses to topography, lead us from archaeology and land-based social history into alternative relationships to place. They can expose the social agendas that have formed the land and bring out multiple readings of places that mean different things to different people at different times ...'

Lippard's statement is bang on: art can affect perception by intellectually and emotionally engaging people, revealing the relational complexities of an issue and inspiring the idea of alternate histories and trajectories. Using mediums like installation and performance to communicate the long history of human interdependence with the city's waterways and to immerse people in the tactile pleasures of water, several Toronto art projects are heightening our aqueous awareness, be it in wading pools, on the banks of an ancient shoreline, along gurgling rivulets winding through city ravines or above the rivers we submerged and turned into sewage tunnels a century ago.

In preparation for this piece, we spoke with people working on a variety of these collaborative projects. Some of them – Human River, Black Creek United, *The Ballad of Garrison Creek* and Stream of Dreams – have been explicit in their mandate to raise

*The Toronto Public Space Committee's Human River spills over the Harbord Street bridge.*

awareness of the relationships between people and water in the city. For others like Jumblies and wade, that outcome was a welcome secondary effect of artistic processes rooted in a particular place. In spite of their sometimes divergent approaches, all of these projects invite urban dwellers to overcome their disconnection from water by playfully exploring and animating this complicated and contradictory relationship. As wade co-founder Christie Pearson puts it, 'Play and pleasure are ways in to accessing the imagination of people in a way that would make them feel more loving toward their environment, more loving toward their city.'

### HUMAN RIVER

In October 2005, we were swept up by a blue procession along the city's fragmented Garrison Creek ravine. This, the Toronto Public Space Committee's first Human River, was a lighthearted re-embodiment of the lost creek brought to life by bodies draped in long ripples of shiny sapphire fabric and children bobbing fish stick-puppets above the ten-dozen–strong crowd. University of Toronto professor Bernd Baldus told place-based stories that spanned 12,000 years, spurring contemplation of the remnants of two bridges obscured in now-filled valleys and the tilting houses and willow trees that indicate underground water. We found Toronto's largest lost river to be a provocative illustration of how our choices to control the flow of water have ultimately displaced us and our watersheds and impeded the direct experience of our sources of sustenance. By joining this Human River, we saw an opportunity to overcome this disconnection from water, and we relished walking/performing the meandering route of the buried creek. Our understanding of the city's environmental history was radically altered; we began to reimagine our relationships to the earth and to each other while we moved in this unfamiliar way. The event made both the possibility of daylighting a watercourse through downtown and of creating a city that coexists symbiotically with nature suddenly palpable.

Since this transformative experience, we have both become enthusiastic organizers of the now-annual event, along with long-time Toronto Public Space Committee organizer Erin Wood. Telling stories through a variety of artistic mediums, recent performances have included clowns digging out the buried Harbord Street bridge with plastic beach shovels, a flock of ghost birds emerging from Stanley Park, water walkers circling in a mesmerizing dance, Spirit Wind drum songs that express the significance of the landscape to indigenous peoples, and artists leading workshops in public schools and community centres along the Garrison. Holding handmade drums, shakers and bassoons, we ripple through remnants of the historical Garrison Creek watershed each October, becoming, once again, a Human River. We are hooked by this chance to savour sensory shifts while disrupting our usual grid-like paths through the city.

*Black Creek United participants chalk maps to sounds heard along the river.*

#### BLACK CREEK STORYTELLING PARADE AND BLACK CREEK UNITED

The excitement generated by Human River has been contagious, quickly spreading to other neglected urban waterways. During the spring of 2006, a group of York University graduate students equipped with percussion instruments and sidewalk chalk dressed themselves in blue and, tracing an underground stormwater pipe, performed one of the many tributaries of Black Creek buried under their campus. Named the Black Creek Storytelling Parade, the event unearthed significant stories concealed in the landscape that are not usually communicated within the university: an unsettled land claim, a Handenausanee village buried under a hydro line just south of campus, the creek as a cultural divide between two neighbourhoods.

After performing it numerous times for campus groups and local high school students, two of the BCSP's initiators were hired by the Art Gallery of York University (AGYU) to collaboratively adapt the ever-evolving project. Laura Reinsborough and Liz Forsberg, co-author of this essay, connected with a dozen seven- to twenty-one-year-olds at the Spot, a community drop-in centre located on the other side of the creek, in Yorkgate Mall. Two months of biweekly workshops culminated in the group leading an arts-based walk through the Black Creek Parkland. The walk, named Black Creek United by the youth, featured an exhibit of photos hung from the low-lying sumac trees that line the ravine, an installation of papier mâché creek birds past and present and a set of activities that invited participants to create their own environmental art and sidewalk chalk interventions in response to the local soundscape. The result was a playful reconnection to the creek, which still meanders by the participants' homes – a place familiar to neighbourhood joggers and dog-walkers, but not typically a stomping ground for children and youth.

On the suggestion of one of the youth artists, the project morphed into a travelling photo exhibit, moving from the Spot to the AGYU and back to York Woods Library. 'It was important that we crossed the creek and consider these two very segregated communities: a privileged academic community that is completely isolated, that has a mostly commuter population, that doesn't know its neighbours, and then also a very stigmatized and racialized community of Jane and Finch . . . and to see how the creek could become a meeting place, rather than a divider for this,' explains Reinsborough.

*Clay and Paper Theatre evokes the Amazon goddess in* The Ballad of Garrison Creek.

The banks of the Black Creek will buzz with artistic activity again as the project continues to respond to requests from institutional supporters and interested youth groups like those at the Jane and Finch Boys and Girls Club.

#### THE BALLAD OF GARRISON CREEK

In the spring of 1998, several years before Human River catalyzed us, controversy over the viability of restoring parts of the Garrison, as proposed in James Brown and Kim Storey's Garrison Creek Demonstration Project, was climaxing. At the same time, Clay and Paper Theatre – Dufferin Grove Park's resident theatre company – was developing its annual play, and David Anderson, the company's artistic director, contributed a response to the brouhaha: *The Ballad of Garrison Creek*. In the tradition of environmental theatre, the intervention used the park as its backdrop; the story unfolded from the point of view of the goddess of the creek, who, as Anderson describes, 'was mightily pissed off at having been turned into a sewage pipe.' Large block-printed banners of trees and water denoted the staging ground, while signs with the word *water* in fifteen languages drew in curious passersby from the park. 'We wanted to look at our attitude toward water,' Anderson explains. Their goal was to encourage the city to deal with lake-water pollution

*A few of the hundred fish created as part of Stream of Dreams.*

at its source, something that the then-estimated $52 million Western Beaches Storage Tunnel project, a 3.7 kilometre system of underground tanks along the lakeshore, was not designed to do. 'We were hoping to lambaste the hubris of the big-ticket approach and to laugh at our own follies, too, while we produced a piece of exciting outdoor theatre,' Anderson adds. The response was enthusiastic, resulting in invitations to restage the play that August in three other green spaces above or near the former watercourse: Trinity Bellwoods, Christie Pits and Little Norway parks. *The Ballad* did not result in daylighting any parts of the Garrison Creek but, as Anderson reflects, 'it did raise a challenge to the standard industrial solution and encouraged people to think of alternatives.'

## STREAM OF DREAMS

That same year, the need to link storm drains to the living habitats into which they discharge sparked the B.C.-based Stream of Dreams Murals Society, and their catchphrase, 'The Creek Under the Street.' This arts-based watershed education project cascaded across the province, and, in the fall of 2007, sprang up in Toronto. Today, there are two schools on Ossington north of College: one is Dewson Street Junior Public School, a typical 1960s building, while the other is a school of colourfully painted wooden fish, mounted to twist and twirl in an imaginary current along Dewson's tall chain-link fence.

A parent-led initiative coordinated by Paola Giavedoni brought the project here. Stream of Dreams called upon students from grades K to 6, their parents and neighbouring residents to pick up a paintbrush, but only after had they learned about local and endangered freshwater fish species, and had been shown area watershed maps, past and present. Most didn't know that Dewson School was founded near what once was Dewson Stream, a tributary of the lost Garrison Creek. Program facilitators traced the urban water cycle from storm sewer to Lake Ontario and back again, emphasizing fish as important health indicators of their ecosystem.

After the program was completed, hundreds of dream-fish were affixed to the fence, given over to the larger collaborative project in an act of release meant to show how combined efforts can yield profound and lasting change. Dewson's Stream of Dreams acts as a permanent reminder of the presence of the tributary and will hopefully spawn other projects.

### JUMBLIES

*Jumblies brings glacial Lake Iroquois to the stage in* Once a Shoreline.

In 2001, Jumblies theatre company began a four-year residency at the Davenport-Perth Neighbourhood Centre, where watersheds and bodies of water seeped into their work. Weekly artmaking workshops drew participants from the centre's numerous social and health programs, such as settlement and literacy, and culminated in two community plays that explored local histories and issues.

*More or the Magic Fish*, produced at the end of the first year of the residency, was a cautionary tale about buried water based on the old folk tale *The Fisherman and His Wife*. Artistic director Ruth Howard can't recall how the theme came about, but remembers thinking it was a good metaphor for the potential to unleash new dreams. This theme of buried water became more meaningful later on when, while doing research for their final production, *Once a Shoreline*, Howard discovered that the community centre's Davenport location was adjacent to the ancient shoreline of Lake Iroquois. The retreating lake inspired her to further investigate themes of memory, travel, fear and courage in stories gathered from participants over the four-year residency. Once completed, *Once a Shoreline* featured music and sound art to conjure Lake Iroquois, passages of recorded factual narrative about its formation, choruses of lake dancers to evoke the fluid movements of water and textured costumes, fabrics, props and puppets in various shades of blue-grey. The audience stood in the centre of the performance space, 'the waters in which the story would unfold,' according to Howard. The result was complete sensory immersion for the audience. Participants, meanwhile, gained a profound sense of the environmental history of their shared landscape. Most importantly, Howard notes, 'all those people who have been washed onto the same shoreline by circumstances mixed and got along together in our ongoing arts project [more] than they did at the start. This is striking. And the project still exists all these years later, floating in that imaginary lake . . . '

### WADE

Since 2004, Christie Pearson and Sandra Rechico have co-curated wade, a festival of immersive performance and art installations in wading pools in the area bounded by Sorauren Avenue, Greenwood Avenue, Dupont Street and Toronto Island's Gibraltar Point. The project started when Pearson

*Michael Caines and Leah Decter invited 'Everyone in the Pool' for felt-making at* WADE.

became so invigorated by her own experience of creating installation art in her neighbourhood wading pool that she approached Rechico about organizing a broader event. The dozens of vivid re-imaginings since have evoked all the senses, and have provoked a cross-section of Torontonians to get their feet wet in projects like Nick Tobier's Busby Berkeley–inspired human-generated fountain at Withrow Park, Michael Caines and Leah Decter's invitation to stomp and stir to make felt in Carr Street's pool and Shannon McMullen and Fabian Winkler's array of water-wave to soundwave converter buoys set adrift at Dufferin Grove. '[Pools] are surrogates for the lakes and the ponds and the rivers . . . We want to be living around living water, and want to relate to living water, but in the city it's very hard for most people; they don't have a chance,' says Pearson. wade provides a rich and unparalleled opportunity to engage with the vital element, even for people who don't have children and wouldn't otherwise think to converge upon these pools. And for Pearson and Rechico, wade's diverse and open-ended programming is an ongoing attempt to answer a driving question: 'How can wading pools become more like living water?'

### RIPPLES

Engagement in these localized artistic processes has had a profound impact on participants and coordinators alike. The act of walking the undulating route of the Garrison Creek undid years of sedimented understanding of downtown Toronto as a flat urban landscape. Our coming to know one lost river kinaesthetically brought about an awareness of all the other rivers and ravines – buried and unburied – that form our city. David Anderson echoes this sentiment when he contemplates how his relationship with the Garrison changed over the course of *The Ballad*'s performance. 'I think it gave all of us a realization that Toronto is a city of creeks flowing into Lake Ontario, that our buried creek is one of those.'

New-found senses of place are the legacies of Dewson School's Stream of Dreams project: 500 children can now tell people what their water address is, and that there is essentially 'a creek under every street.' This expanded water awareness also translated into very practical matters like the children's use of non-toxic paints and their conservation of water during painting and cleanup.

Working along the ancient shoreline of Lake Iroquois was a transformative experience for Ruth Howard too. 'Every time I go up or down that hill, I think of the lake,' she says. 'It's very much a part of my sense of place now, and before it entirely wasn't.' And for Christie Pearson, wade brought together what she saw as little lakes scattered across the city. Over the course of the weekend event, wade festival-goers can 'walk or bike from pool to pool, and get a sense of the artificial networks of water underneath, and the system of parks and open spaces that make something bigger than people think,' she explains.

So, what would a future set afloat by creative and conscious connections to water be like? It would be an ongoing commitment to commemorate water in Toronto with a multiplicity of traditions, performances and festivities that reflect our social and environmental diversity. Each of the city's rivers, buried or intact, could have its own annual celebratory walk on dates significant to each watershed. On other special days, like Canadian Rivers Day and World Water Day, these waters could converge in a city-wide procession that would highlight the connections between each day's commitments to action. Meeting together along the shorelines of all five Great Lakes, we could take part in a giant gathering to honour the Mother Earth Water Walk, the successful circumnavigation of these sacred waters between April 21, 2003, and May 12, 2008, by Anishinabe elders who took turns carrying an eight-litre bucket of water as they made their way around a different lake each spring to symbolize that we all must mindfully carry water for generations to come.

Tributes to tributaries could happen more frequently and bring smaller communities together to mark full moons, spring melts and winter freezes. In the summer, we could reinstall ancient shorelines with waves of water, light, sound and movement that lap up where they have been absent for hundreds, even thousands, of years. We could make sculptures in our public spaces that collect water and erupt into polyphonic fountains during a rainstorm and melt-forms during a thaw. The buoyant outcome of this re-immersion in multi-sensory community-based art would be an increased appreciation of the social, cultural and environmental importance of our lovely Lake Ontario and all the watercourses that feed into it.

Maggie Helwig

# Downward

There is a part of the collective Toronto psyche – we have to admit it – that wants to be New York, that wants to be big and bright and metallic, thrusting upward, meeting our fate, for good and bad, far up in the skies, the realm of shining towers and airplanes.

But we are not like that, finally, the CN Tower more of a mild embarrassment than anything else; we're a city that travels downward, a subtle city, half-buried, and our fears and our connections happen below ground level. We think of subway crashes, and strangers pushing us onto the tracks. When SARS was in Toronto (and, in reality, being transmitted mostly in hospitals), it was on the subway that people wore masks; it was subway seats and poles and escalator railings that the newspapers analyzed as infection vectors.

We have many subterranean spaces in Toronto, and in our minds. A lot of cities have a subway; not many have anything like our PATH system, those labyrinthine interconnected kilometres of commerce, perfectly removed from any sense of time or direction, all underground fountains and food courts and Laura Secord for miles. There would seem to be no good reason to be in the PATH system; few of the stores there are unusual or worth seeking out, and the downtown core is, in fact, compact and easy to navigate above ground. But I have never been in the PATH and found it empty. There are always people walking in the PATH, even when the stores are closed. We wander there, disoriented, vaguely companionable, wondering where we are.

Our literature, too, goes underground. Our most thoroughly mythologized public building is a water-treatment plant and its buried system of pipes. And while perhaps the single most iconic Toronto scene in literature is the fall of the nun from the unfinished Bloor Viaduct in Michael Ondaatje's *In the Skin of a Lion*, it may be that the true meaning of that scene resides not in the vaulting bridge but in the space beneath it, over which the characters hang suspended. The Don Valley, that deep, carved, green rivercourse that slices the city into two only partially reconciled parts. (West end, east end – which one do you live in? How often do you travel to the other? How often without a sense that you are entering a foreign land?)

Any novel that wants to be a Toronto novel has to reckon with the ravines; in our imaginations, even more than in reality, they are the city's dominant shape. It may be too easy, especially after Margaret Atwood's *Cat's Eye*, to use the ravines as the repository of buried childhood trauma, but writers do keep doing it. And other shapes – Dionne Brand's *What We All Long For*, which begins with the characters poised above the Humber Valley in a subway train, a subway that is not underground but in the air, like the falling nun, above the great depth of a valley, the strange contradictions of our particular urban space.

Water shapes our social world; the locations of cities, their shapes, are determined almost entirely by water, and Toronto would not be what it is if Lake Ontario were not a lake of a particular shape, in a particular place, as London is shaped by the Thames and Rio by the Atlantic. But our relationship to our rivers and their ravines is a bit different. We have not completely filled them in or built over them, but neither have we openly shaped the city to their demands. We live with our sunken rivers in a quiet unspoken collaboration, changing their courses sometimes, dumping things out of sight, but letting them continue, sometimes going back to them, cleaning up, trying to erase some of our presence.

And sometimes we simply go downward.

At the edges of the ravines, on the high ground, are some of the most privileged neighbourhoods in Toronto – Rosedale, Forest Hill, the Bridle Path – and further north, expanses of golf and country clubs. But if you walk down the hills, though you enter a wilder space of bloodroot and fern and poison ivy, you do not leave the social world. The ravines are inhabited as well, and directly downward from Rosedale are other members of the city; the people, mostly men, who live in tents and boxes,

who mainly do not want to be found, the sharp end of deprivation in a wealthy city.

A photograph in winter: the street nurses make their way up a snowy slope of the Rosedale ravine, carefully, sleeping bags under their arms, the emissaries of the comfortable world, toward an improvised tent, toward people who live without comfort.

Another photograph: a very thin, aged man, surrounded by piles of plywood, his few possessions, a notebook in his lap, on

the bank of the Don. His name is Fred. 'I'm not homeless,' he says, fists clenched in determination. 'This is my home.'

And another: Peter, a younger man, looking nervously out from a dwelling he has built from branches and covered with leaves, hidden, safe.

Many things have brought these men here. For all of them, probably, there is some desire to escape, to hide from the city, from the world of other people. And yet they are here, they are part of our ecosystem. We may see their tents when the subway

passes over the Don. They may be fathers or brothers or sons who are lost; forgotten or remembered. There is someone or something that each of them loves.

Sometimes I think this is how we deal with conflict in Toronto, how we deal with the great gulfs of privilege in our shared space. We do not confront each other; we do not drive people out, explicitly, or fight for territory. We are polite, restrained, quiet. We live side by side and yet hide our faces from each other; we allow people to go downward and find their shelter there. And they also choose, in some sense, to go down, to go out of our sight, to find a place by the side of the buried waters.

Desire, too, travels downward, and in this way, too, the ravines are part of our social body. If I think first of the hidden people in their tents and lean-tos, there are others who know the ravines especially as a cruising area, where men encounter each other for concealed and sudden passion, sex without narrative, by water and fallen trees. Or simply the paths where Rosedale teenagers drink vodka stolen from their parents' liquor cabinets and make out, daring in their own minds. Joggers in their bright suits move up and down the trails in the mornings, their own visions of their better selves in front of them; obsessive naturalists with night-vision goggles come out in the darkness. What we want the most is always something we partially hide, something to do with secrecy and discovery, this concealed space beneath the trees, the graceful houses poised above it, the light from their windows almost reaching us.

The trails of human desire in all its strange forms reach into every part of the city, even the overgrown valleys; and the city's other inhabitants wake in the spring and travel up from the ravines, raccoons building their nests in our houses, falcons plunging and dropping among the high towers of Bay Street.

We are not apart from them. We move downward, we go to the rivers. We all need this water to live.

Michael Cook

# Water underground: Exploring Toronto's sewers and drains

The first Toronto drain I ever entered began in a park on the Vaughan side of Steeles Avenue. A cage of steel bars covered the inlet, designed to keep flotsam and children out of the darkness beyond, its purpose undermined by an unsecured, free-swinging door. I crouched down and slid inside, discovering not much more than mud, low ceilings and a lovely little space fifty metres downstream where seeds had sprouted in sediments deposited beneath a large metal grille. Some of the plants stood nearly a metre tall in the waning, late September light, having found by accident of wind or water a tenuous compromise to the manicured lawns that prevailed on the grounds above. Just beyond lay the drain's outfall and a slow, nameless tributary of the West Don. It was a beginning, albeit a humble one.

The observation that, as a public, we don't think enough about the mechanics and consequences of storm and wastewater handling has probably reached the level of cliché, yet the essential truth remains intact: it is easy not to concern ourselves with things that are buried out of sight and mind. That said, I am not going to write about the failures of our stormwater management systems, at least not directly – if you're reading this volume, you probably know a little about downspout disconnection and porous surfaces and E. coli. Instead, I'm going to try to convey what it's like to experience the systems themselves, and hopefully along the way I'll share a few interesting facts about how we as a city came to be where we are today.

## UNDERWORLD DREAMS

By the time I was stumbling through that muddy culvert between Steeles Avenue and Fisherville Road, I had already found my way into at least a dozen more glamorous tunnels in Burlington and Hamilton. All served the same purpose, though, carrying creeks and rainwater underground, always to facilitate urban development. Having cut my teeth exploring the steam tunnels beneath my university campus, it was only natural, when I found myself following fragmented creeks through western subdivisions and highway interchanges, for me to don headlamp and waders, check the weather forecast to make sure it wasn't about to rain, and press on to see where the water went when each creek almost inescapably dipped below grade.

It would take several more years and dozens more drains and sewers before I really understood the full dimensions of what I was seeing and of what existed underground. At first everything I found was located by following visible watercourses to points where they disappeared; more recently, a scattering of old maps from long-forgotten reports have helped open up the otherwise invisible sewer systems of Toronto's core. Not every tunnel has been my own find, of course – I've been aided immensely by collaboration with a small number of colleagues from Hamilton, Toronto, Montreal and further afield. One is a baggage handler, another a construction plumber, another a professional photographer, another a dance instructor – what we all share is an inquisitive fascination with the watersheds that have been buried beneath our feet.

#### PEERING INTO DARKNESS

There's a manhole in the middle of Christie Pits Park, near the field house. When it's not plugged with a large plastic stopper, it's perhaps the most perfect place to catch a momentary glimpse of what lies beneath the city's surfaces. Until the recent installation of this isolation device (to prevent runoff from the park entering the sewer, or perhaps just to collect it for study), the manhole's decades-old collar and lid had been slowly overrun by the advancing turf such that, though it was no longer an effective access point to the sewer below – the ladder irons having rotted away – it remained a window into the earth. If you got close enough, you could hear the roiling wash of grey water that for the last hundred years has traced the path of the most storied of our city's lost waterways.

Garrison Creek once flowed in a ravine here that, quarried and deforested in the nineteenth century, became the pit we know today. Downstream, the former ravine lands became other parks: Bickford, Harbord, Trinity Bellwoods. Damaged and pressed into service as a public sewer, the creek was forced entirely underground before the onset of the Great War, first into brick pipes and arched concrete tunnels, and later into modern concrete conduits constructed in the 1960s to relieve the burgeoning city's pressure on the original system. Not even burial could end the creek's substantial surface impacts, though – across Bloor Street from the pits, the footprint of the public school board's Bickford Centre, built in 1965, is cut in two to accommodate the sewer conduits passing beneath it. Spot one of the many signs installed at street corners in recent years to pay tribute to the lost creek, and it's a fair bet the old brick sewer isn't far away, quite possibly right underneath you. Though it's invisible on the ground, in satellite photographs you can actually see the surface shadow of the Garrison sewer laid out like the remnants of a neolithic track across the west side of Christie Pits and down through Bickford Park toward the buried bridge that a century ago spanned the creek at Harbord Street. The effect of the infrastructure on surface-water drainage in the parks probably accounts for its visibility from the air.

Beneath the manhole, the tunnel is a concrete arch just under two metres tall. When it isn't raining, the effluent from the West Annex's showers, toilets and laundromats is largely contained in a half-metre-wide trench in the floor of the sewer. In places, groundwater trickles in through seams in the concrete or drips from the ceiling. Just downstream, smaller pipes bring in more flow from points east and west along Bloor Street, and the shape of the tunnel changes where it was rebuilt in the 1960s to accommodate subway and school board. Here the air is thick with humidity and the flow rushes down a short slide before slowing behind a diversion weir. The tunnel occasionally rumbles as trains pass in the adjacent Bloor-Danforth subway, while the metallic clang of cars driving over manhole lids in nearby streets offers a constant reminder that we are in the underworld.

Above the pooling sewage, a large dry pipe in the eastern wall leads down into what may be the original Garrison Creek sewer south of Bickford. The old conduit at the end of this overflow structure is a round pipe five feet in diameter and built of yellow and, further south, red brick that is now semi-derelict and does not appear on the maps I've seen. It does receive enough organic matter from local sewers to support a number of rats, and to occasionally flow into other elements of the Garrison sewer network further to the south.

Other shafts and other portals offer similarly apt projections of the world beneath them. The sidewalk that slants southwest from the Rosedale subway station back toward Yonge Street and Aylmer Avenue is downwind of an often odorous lid that leads into another yellow-bricked conduit, the Rosedale Creek Sewer. Stop and listen, and you can hear the flushings of

Yorkville and southern Forest Hill tumbling into the ravine that now holds the Rosedale Valley Road.

The walls of this pipe have wizened with the infiltration of groundwater and tree roots. Underground drips and springs deliver quantities of calcium and other minerals that have, over decades, become quite significant, and great fossilized streams now reach down from each porous moment of the aging mortar that holds the round brick walls together. Tree roots require larger portals; they appear in great shoulders of moisture-suckling growth from a handful of side connections and manhole shafts, and typically bear their own deposits of calcium, flowstone and dried effluvium. As this tunnel passes through Ramsden Park, it slides down a series of long, shallow ramps, each one accompanied by a metal railing to permit workers to safely negotiate each slippery hazard. East of Yonge Street, a more abrupt fall exists – here the flow rushes down what may be a steep staircase. We have yet to push past this point.

This sewage is intercepted and sent to the treatment plant at Ashbridges Bay, but an overflow conduit beyond the diversion plunges into the Don River; built in the 1930s, it is the tunnel pictured on the cover of one edition of Michael Ondaatje's novel *In the Skin of a Lion*. Again, this was all once a living creek; referred to variously as Castle Frank Brook or Rosedale Creek, there are photographs at the City of Toronto Archives of the footbridges that once crossed its waters in the parks west of Yonge Street.

#### BURIAL GROUNDS

The floor of a well-vegetated ravine seems like an unusual place to find manholes, but since we've used ravines as corridors for many of our modern trunk sanitary sewers, observant readers will know that the floors of our valleys are lousy with lids. While early creek sewers were a belated public health response to the already established use of the creek as a sewer, by the 1960s, we were removing creeks from the city's landscape more for reasons of expediency – they occupied key real estate, had an inconvenient habit of occasionally flooding the land around them and, once encased in concrete, made very effective and invisible sinks for stormwater that collected in smaller sewers beneath the former creek's watershed or further afield.

The Cedarvale and Nordheimer ravines offer a prime example of this strategy. As part of the failed Spadina Expressway project, Metro Toronto funded the construction of several large-diameter storm trunk sewers to provide drainage for the sunken highway from Wilson Heights south to Davenport. This was a particularly important boon for the then Township of York, which was grappling with a pressing shortage of sewer capacity. While the more northerly of these storm trunk sewers stretched east to the Don and west to Black Creek to drain what ultimately became Allen Road, the largest was laid beneath the Cedarvale and Nordheimer ravines, burying the Yellow Creek and the remaining open-air extent of Castle Frank Brook and converting the valleys into a natural extension of the trench that was being dug north of Eglinton. Beneath both the anticipated highway and the massive Spadina Storm Trunk Sewer, the Spadina subway line would be installed in tunnels bored below the ravines.

Ironically, a few years after this trunk storm sewer and another near Glencairn helped solve the sewerage crisis in the eastern portion of York Township, York Council would emerge as a key participant in the opposition movement that ultimately forced the expressway's construction to end at Eglinton Avenue, just above the ravines. Their main objection to the road? It would pave over parkland in the Cedarvale Ravine that was essential to the community. That the extent and accessibility of this parkland had only recently been realized through the sequestration of the ravine's former creek into the new highway's key drain appears to have received little comment.

The Spadina Storm Trunk Sewer stretches from Eglinton Avenue southeast to an outfall in the Park Drive Reservation lands east of Mount Pleasant Road. Up to four metres in diameter and running for close to six kilometres, it is one of Toronto's largest and longest sewers. Inside, about a dozen artificial waterfalls carry collected stormwater down the creek's former path. Each waterfall is serviced with catwalks and large plunge pools, fascinating architecture that few Torontonians have ever seen. This enormous conduit twists repeatedly to follow the uneven path of the ravines it shadows; hiking its full length from the Park Drive outfall is an endurance test in rubber waders and disorienting when, the first time you emerge through a manhole, you find yourself on the floor of the Cedarvale Ravine, with no immediate idea of where you've ended up.

In the northeast, another section of creek lost in the 1960s nearly escaped burial. Consultants from the engineering firm Gore & Storrie Ltd. recommended that a section of Wilket

Creek north of the 401 remain above ground, albeit relandscaped and reinforced as an open-topped drainage canal. They argued that it would be less expensive to build it this way, and that this design would preserve at least a compromised version of the area's previous landscape. The creek, a major East Don tributary that flowed down a wide, shallow flood plain in central North York, had the misfortune of running through the middle of Willowdale's post-war housing boom. Residents were concerned that the waterway was a danger to children and property values, and their opposition was significant enough to scuttle the open-air plan, and the creek.

Today, Wilket Creek runs just below the surface of its former course, from north of Finch Avenue southeast to York Mills Road at Bayview. There it finally emerges to flow southward, adjacent to the Bridle Path and Edwards Gardens, before emptying into the West Don near Eglinton Avenue. The tunnel, which grows to nearly five metres in diameter, occasionally opens up to the surface via large steel grates similar to the one sustaining plant life in the culvert below Fisherville. In this case, the main beneficiaries of the sunlight are colonies of green algae that survive on the exposed walls adjacent to these openings. In the wintertime, snow occasionally accumulates on the ledges beneath these skylights – the fact that the drain is so open and proximate to the surface keeps its ambient temperature much closer to what's overhead than most other systems, where conditions are considerably warmer, even in the depths of winter. In fact, this particular storm trunk is so shallow that some sections are apparent from above ground as a low berm set toward the western side of the flood plain, its panelled concrete walls concealed beneath less than a metre of earth.

### EXPERIENCING WATERSHEDS UNDERGROUND

We normally read the city in the straight lines of pavement and surface travel: the major street nearest to where we live (whether it runs north/south or east/west), the right angles of most bus-to-subway transfers, the imaginary diagonal that serves as an impossible hypotenuse linking our present position with our intended destination. While places within the city are grounded in a physical environment, the landscape in which they are fixed is often more abstract than it is tangible.

In charting and perpetuating natural watersheds, the sewers below Toronto maintain the pre-urban landscape even in places where we have succeeded in wiping most of its traces from the surface. Walking a major sewer, we follow the engineered ghost images of the city's former creeks and streams, and experience a succession of neighbourhoods in integrated swaths of runoff and effluent. It's like walking into those diagrams of the water cycle that we carefully coloured and labelled in elementary school geography; here below the city, the watershed exists in something approaching its original continuity, and in architecture cavernous enough to hike.

Above ground, the Fashion District, Kensington Market and the eastern Annex are most often imagined as wholly separate places intercut by major streets, institutions and anonymous residential blocks. Yet runoff from their roads and alleys all drains into the same trunk sewer, a mammoth conduit begun lakeside in 1965 and mined out beneath the existing city over the course of ten long years. Its official name – the Garrison Creek East Branch Storm Trunk Sewer – points to the fact that even the City engineers of the day had forgotten the waterway whose drainage basin the tunnel largely served: Russell Creek, Garrison's discreet neighbour to the east. As the trunk drain drives southward twelve metres below street level from its beginnings in the vicinity of Davenport Road at Bathurst Street, it takes in storm flow from major tributary pipes at Lowther Avenue, Bloor, Harbord, College, Dundas and so on until it has grown to nearly four metres in diameter. Passing Front Street, the water washes without ceremony into a flooded siphon that takes it beneath the rail lands and on into Lake Ontario; hundreds of young lake fish congregate here in the small pool at the top of the siphon, safe from avian predators.

In an area of the city where the rare and distant echo of water running below the streets is the only reminder that it once flowed freely across the land here, walking the length of this trunk sewer is a revelation. The same holds true elsewhere – the earth below Rosedale and Forest Hill is riven with pipes and shallow box drains that replace streams that once washed across the eastern beaches, and the light industrial economy of the amalgamated city churns above buried creeks on the fringes of North York and Etobicoke. The water is still there, locked into an institutional obscurity by our own lack of interest and the opacity of public works. We're encouraged to treat infrastructure as a black box, something that just happens, while this decade's increased security concerns have only added their own layer of murk. Along the way, we miss some fascinating opportunities to be conscious of and live within the ways that the land and this city functions. While wading in full-bore with rubber boots and lights has been my and my colleagues' preferred method of experiencing Toronto's water, there are other more socially acceptable and lower-risk approaches.

Turned back somewhere beneath Bickford Park by high water, we eventually regain the Garrison Sewer further south. Here, a relief sewer was constructed sometime after the original brick pipe – a concrete arch three to four metres tall, floored in brick, that carries the storm overflows of the other Garrison sewers from a large chamber north of Dundas Street south to the Strachan Avenue storage tank of the massive Western Beaches Storage Tunnel. For six hours we breathe the mists and listen to the babbling voices of a waterway that was eulogized as 'the pretty, purling brook' in a newspaper clipping, undated but likely from the late nineteenth century, that was reproduced in a 1998 Toronto District School Board publication.

All manner of pipes now feed into the relief sewer: an ancient, balloon-shaped brick conduit perhaps a metre high; more modern concrete pipes whose trickling, colourful fluids hint at upstream overflows; a small opening gushing combined sewage even after several days of dry weather. As we make our way downstream, the tunnel reverberates with the sound of trucks driving over manhole covers, and the force each passing streetcar transmits into the earth. Somewhere near the Princes' Gates, the sewage pools on all sides of a pair of low dams – in front of them, the flow slowly drains into what is probably the last stretch of the Exhibition Interceptor, while the deep, stagnant water beyond the second dam alludes to an unseen control structure for the newer storage tunnel.

It may not be the postglacial landscape of the original creek, but this underground geography is no less laden with complexity and heritage. After walking the insides of combined systems like the Garrison Sewer, there can be no doubt that more naturalized approaches to stormwater management, like the network of surface ponds that were proposed for the Garrison watershed in James Brown and Kim Storey's Garrison Creek Demonstration Project, are desperately needed. Our urban watersheds as they exist today are at once overbuilt yet still inadequate for dealing with extreme storm events. They are also almost completely inaccessible without the transgressive approach that my colleagues and I use to reach them. While much money, ink and time has been expended on improving access to the lakefront at the foot of the city, the water that is everywhere beneath our feet is no less deserving of our attention.

Exploring these sewers resuscitates the existing underground landscape from the abstract treatment it receives in environmental proposals and daily life. Toronto's sewers are real places, and they are worth knowing in all their cryptic, three-dimensional glory. We should no more write this infrastructure off completely in pursuit of a greener future than simply accept the invisible and untenable status quo. For reasons of cost and practicality, this city will always have trunk sewers where creeks once flowed, but celebrating the most fascinating of our subterranean streams elevates the whole of the watershed in our general consciousness. With a greater awareness of the shape, structures and history of our sewer networks, the task of building the collective energy and political will necessary to begin daylighting portions of our buried watersheds may become much easier. That awareness starts with taking the time to peer through the openings of a few manhole lids. While I wouldn't go out of my way to encourage people to attempt what my colleagues and I are doing, it won't hurt anyone to take a look down a grate, to listen for the water in their neighbourhood and to ask that our public works do more to tell and *show* us what's down there. The water is there, and it is up to all of us to bring it into the light.

FOLLOWING PAGES

193
*Rosedale Creek Sewer.*

194
*Garrison Creek Relief Sewer.*

195
*Typical section of suburban storm sewer.*

196
*Abandoned stone culvert beneath the West Toronto Junction.*

197
*Cowan Avenue Storm Trunk Sewer (at Bloor Subway).*

198
*Highway 427 Storm Sewer.*

199 TOP
*Garrison Creek East Branch Storm Trunk Sewer.*

199 BOTTOM
*Storm sewer beneath West Toronto Junction.*

200
*Wilket Creek Storm Trunk Sewer.*

# Directions

RiverSides
5 THINGS YOU CAN DO FOR YOUR RIVER
Your home—Your river
make the connection!
THE PATH TO CLEAN WATER BEGINS IN YOUR BACKYARD
This guide illustrates actions you can take at home to prevent stormwater pollution and improve the quality of our rivers and lakes.

DOG POO FROM YOUR LAWN ENDS UP IN THE LAKE.
PLEASE STOOP AND SCOOP.
TORONTO
toronto.ca/water

James Brown & Kim Storey

# Buried alive: Garrison Creek as a rediscovered extended waterfront

In 1991, our firm, Brown and Storey Architects, completed a study for the City of Toronto called *The Open Spaces of Toronto: A Typological Classification*, which categorized seven types of public open space in the city, compared their characteristics to similar international examples and identified the systems that connect these open spaces. One type of linkage in Toronto stood out for us: the network of buried creeks that underlie many of the city's parks and schoolyards, and the curving streets that run against the normal grid. It seemed to us that this was a fundamental characteristic unique to Toronto and was worth pursuing further, so we carried our research forward to look at the remains of the Garrison Creek ravine in particular, and began an education that has encompassed ideas of exploratory mapping, public space, the evolution of the city and the role of infrastructure in regenerating whole urban precincts [1].

A fascinating consequence of our work has been sharing with the people who have lived along the invisible banks of the creek's buried ravine over the last century their memories of standing on bridges now long-buried, of the constant smell of garbage from successive waves of landfill, of a tributary flowing through a rhubarb patch on their grandparents' farm. The most remarkable story we've heard, however, was from someone who lived next to Trinity Bellwoods Park in the fall of 1954, when Hurricane Hazel hit Toronto. Against the backdrop of the devastation along the Humber River and the extreme flooding of the Don, Garrison Creek, then buried for already seventy years, made its presence known. The severe rainfalls that had swollen the Humber and the Don rivers to their hundred-year-storm levels had also fed into the pervasive and hidden creek landscape of Toronto, and into the underground sewer where the Garrison had been entombed in the 1880s. The extreme pressure of water rushing through the sewer brought Garrison Creek exploding above ground, blowing off manhole covers to create a line of geysers that sprayed high into the air.

That, to us, is a story about a creek being buried alive. And the more we worked, the more we expected to discover a more dynamic relationship – a 'connectedness' – between the natural watershed and the constructed infrastructure of the city, a connectedness that could act as a catalyst for a real transformation in the urban landscape. As a result of our research, water has become much more than a nostalgic memento for us; it is an untapped dynamic that can re-energize our open spaces, and it is a way to recall a more integrated functioning of the landscape.

The best way to describe the research into water in Toronto is with one word: *dry*. After the stories and the old quaint pictures have been collected and the maps have been traced, complacency settles in, usually followed by commemorative signage and bronze letters that stand in for what's not there. These efforts are comfortable, safe and quantifiable, and, after a brief flurry of 'discovery walks' and newspaper articles, we're able to safely bury the creek away in the bottom drawer with the other civic nostalgia. After a reasonable amount of time, the material is rediscovered and the same cycle is repeated.

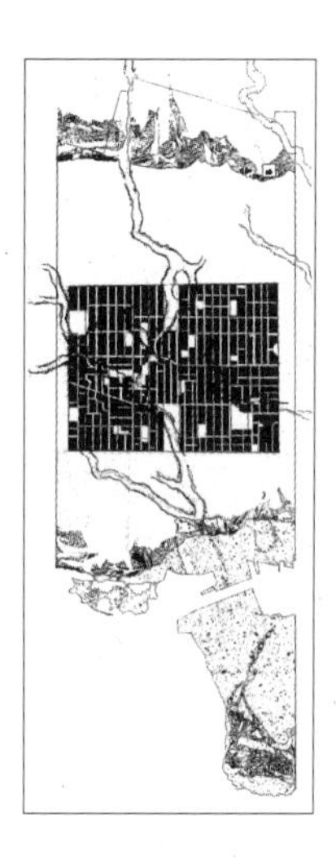

1 *Garrison Creek Watershed: landforms and the city. A working map area was constructed of the Garrison watershed from the Davenport Road ridge to Lake Ontario. The natural landforms of the Davenport Ridge, the original lakeshore and the Garrison Creek ravine imprint are superimposed on the present waterfront profile and highlighted urban grid.*

We need a confluence of design innovation, community support and political will to engender a sea change in how we think of water, a change that will see water become a creative part of our public spaces, and see our parks and schoolyards linked to their original role as part of a natural watershed that knits city neighbourhoods into the Lake Ontario shore. The Garrison is only one of a dozen similar ravines – imagine a repeating series of 'greenways' made up of public squares, schoolyards, parks and boulevards that lead to the waterfront. Imagine if we could use public infrastructure dollars to replace a 130-year-old system of buried pipes with a system that recognizes the natural topography of our watersheds and that would feature a new generation of reinvigorated parks. The real potential of rediscovering our buried creeks is more than annual walks that trace the remains of buried bridges and landforms: it is a major public works project that could redefine the way a city can be fully restored to its natural founding geography. That sort of vision needs political leadership and a fully integrated working relationship of City departments (planning, public works, economic development and parks), the school boards and federal and provincial levels of government.

Garrison Creek and the ravine it flowed through are founding landforms of the original city, their name taken from the siting of the garrison of Fort York at the creek's mouth. Early maps, like the 1818 Philpotts Plan of York and the 1858 Boulton Atlas, show the limits of the new city as the Don River to the east and Garrison Creek to the west. The 1876 *Bird's Eye View of Toronto*, an ambitious aerial view of the young city drawn at a time when the urban and natural landscapes still coexisted, shows early industries like foundries and distilleries located along the creek; at that time, the creek was navigable by boat as far as Bloor Street. Thirteen bridges crossed over it to form an early balance between the growing settlement and the natural ravine, which was over ninety metres wide and twenty metres deep at its widest point [2 3 4]. By the late 1880s, however, the unhealthy odours and appearance of the creek resulting from its use as a dump led to its burial. Garrison Creek became the Garrison Sewer, part of the ambitious layer of wastewater infrastructure constructed by the City in the late nineteenth century that still exists today. The ravine, under pressures of development, eventually disappeared into a scattered collection of intermittent depressions in the city landscape.

2 3 *Two generations of the Crawford Street Bridge, photographed in 1912 and 1915.*

More than 100 years after the creek was buried, we began our initial explorations into the landform of the Garrison Creek Ravine in the more intensive wave of research that followed the *Open Spaces of Toronto* study. We superimposed the large parks that were commonly understood to be part of the Garrison's original path on a reconstructed segment of archival maps that showed the natural watershed and early settlement. We quickly found the seemingly disconnected schoolyards, churches, parks and curving streets, like Crawford and College streets, that depart from the city grid and that form the traces of the buried ravine, and found that far more of them were revealed in archival mappings, photographs and through on-the-ground research. These urban signifiers became part of a growing network that we described as part of a 'co-evolving system' in the urban landscape of Toronto. By co-evolving, we meant that environmental and urban development systems cannot be looked at independently, but rather only in connected pairs – that changes to one system necessarily affect the other.

Taking a normal city map like a basic streets-and-properties plan, we drew up a new collection of maps that focused on the landforms that exist just slightly below the surface, challenging the muteness of the street pattern and showing relationships between the original landform and the superimposed 'park lot' survey of early Toronto.[1] By doing this, we hoped to illustrate how the natural and the constructed have had a continuing

1 Park lots were orthogonal land divisions extending from Queen Street to Bloor Street laid out by Simcoe's engineers in the 1790s as enticements to prospective gentlemen settlers.

conversation in city development. For example, we realized that churches that seemed hidden within neighbourhood streets – St. Anne's on Gladstone Avenue and St. Matthias Anglican on Bellwoods Avenue, for example – were once on the now-vanished banks of the ravine. Likewise, Little Italy is defined by the bend in College Street, a result of early urban planning that followed the lines of the landform. The sequence of Christie Pits, Bickford Park and the Montrose School yard were once a fully continuous open space larger than Trinity Bellwoods Park that was connected to the city streets by bridges. The houses along Shannon Street have two front facades – one that faces onto the street, and another that faces onto the Ossington Old Orchard School yard, or onto the former ravine. The list goes on and on, telling the story of Toronto and Garrison Creek, a story of the rational superimposed on the phenomenal – of the Victorian sensibility that required the most effective and useful division of land at odds with the untamed dynamic of the natural landscape.

4 *The Harbord Street bridge over Bickford Ravine, 1910.*

These discoveries all led to our greatly expanded view of the city as a work in progress. We understood it as a co-evolving system that should be characterized by a continuing search for balance between the urban and natural landscapes. We also saw that a reconfigured relationship between Toronto and its original landforms would be a major step forward for the city because it would provide a new impetus and structural framework for thinking about the relationship between infrastructure and open spaces. As we worked to fully understand the real connection between the original natural watershed and city infrastructure, we began to recognize the enormity of the influence of the Garrison Ravine on a wide range of city characteristics. The archival mappings and photographs we'd studied demonstrated that Garrison Creek had been a magnet not just for open space, but also for cultural, academic and religious institutions, greenhouses and aggregate mining, and that it had been a focus for new settlement in the west end of Toronto. We unearthed late-nineteenth-century advertisements selling new building lots along the 'Garrison Parkway.' (More ominous was the sketch showing a Garrison expressway as an early, unrealized alternative for moving traffic in the 1950s.) Even after the creek had been buried into a sewer, several generations of bridge-building allowed the ravine and the new settlement to coexist as two systems – the urban and the natural – in a balance

[5] *Cultural infrastructure: Aristophanes'* The Frog *being performed in the natural amphitheatre in the Garrison Ravine, Trinity Bellwoods Park, 1915.*

that gave equal importance to each. But, as the city continued to intensify, the ravine became a target for landfill. Bickford Park, once a brickworks site, was filled with garbage. The north end of Trinity Bellwoods was filled with the excavations for the Bloor-Danforth subway line. As the ravine was filled in, the bridges disappeared, with several, like the Crawford Street and Harbord Street bridges, buried intact. The ravine became a collection of fragments, tied invisibly together below the surface by the water still running through Garrison Creek.

We like to call our research into the Garrison Creek ravine a 'regeneration project': fertile ground for revitalizing a precinct of the city. The wealth of influences of the ravine on street patterns, built form, local economy and water infrastructure points to the creek's fully functional integration into the city's network of industries, housing and institutions [5], a coexistence that was thrown out of balance by the developing view of the city as an area of zones – rigidly bounded single-use areas. When we think about the city as a complex series of multi-evolving relationships, we think that improvements to the city can work best by inserting 'patches' that have a cumulative spinoff effect. A fundamental change to its infrastructure, like a connected series of ponds in our parks that simultaneously collect and treat rainwater, can at the same time begin to rejuvenate open spaces that in turn become attractors that increase economic activity, which become centres for environmental education, which become desirable sites for new institutions that bring vitality to neighbourhoods. In the same way that the gradual disappearance of the ravine seemed to be tied to the removal of major institutions (like the relocation of Trinity College from Trinity Bellwoods Park to its present site at Hoskin Avenue, a site that borders Philosopher's Walk, a remnant of the mythic Taddle Creek), the rediscovery of the Garrison in physical and meaningful ways can be used as a potent catalyst in a future composition of Toronto. Instead of staying in the realm of mournful before-and-after photographs, the renewed role of Garrison in the city landscape should be addressed by asking, 'What if?' [6 7]

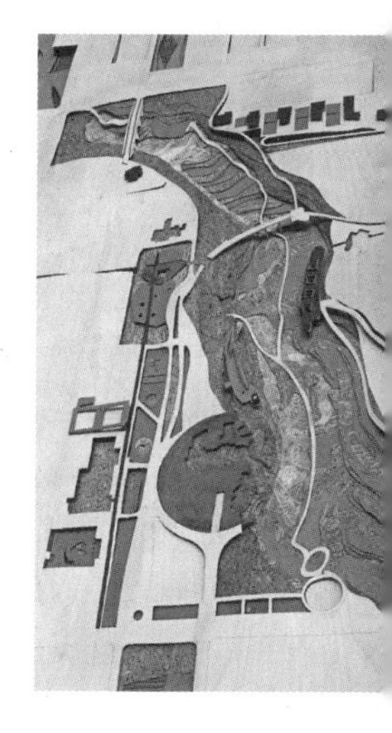

6 Surfaces of loss: *a model of Trinity Bellwoods Park proposal.*

We started our what-if thinking by using maps of the sewer system as a base for a new set of scenarios. We superimposed separated storm sewers onto the open-space network of the traces of the ravine in an attempt to find moments of confluence and possibility, where the intersection of systems of open space, streets and community uses coincided with significant but hidden remains of the creek and its watershed, producing 'hot spots' for urban interventions that would be firmly grounded in the unique character of Toronto's founding landforms. Quickly, the regeneration of the Garrison Creek ravine became a catalyst for an innovative approach to stormwater management, resulting in improvements to both public space and to the quality of the Lake Ontario waters.

For example, the aforementioned network of parks and schoolyards along the Garrison watershed could be the location for a set of natural stormwater-management features that either store stormwater in cisterns or create open rainwater ponds that reuse and recycle water. The dollars invested in a new water infrastructure would also fund new landscape features and public space improvements like fountains, wading pools, water gardens and reflecting ponds. In schoolyards, the presence of these new water features could become part of an environmental-education initiative. With the perennial threat of school closings, the priceless commodity of those public lands needs to be strategically protected. Investing in schoolyards also encourages greater community involvement, which knits the school more firmly to its neighbourhood, creating a custodial, eyes-on-the-schoolyard atmosphere, and an entrenched public claim on the critical relationship between

1 **Original Landscape formed by ancient receding glaciers;**
The creek meandered down to the lakeshore, part of a network of creeks and ravines that score the plain running from the Oakridge Moraine to Lake Ontario. Set in the densely forested plain, the length of the ravine, although forested as well, would have been used as a comparative clearing and way through the natural landscape to the lake.

7 *Co-Evolving System: from 'Rainwater Ponds in the Urban Landscape' by Brown and Storey Architects*

our local schools and our communities. Water-treatment infrastructure is connected to a natural system; we should think of it as having the ability to improve the amenity of public spaces through redesign, of education and community facilities especially. Instead of the standard end-of-pipe solution, the money that has been invested in a single point on the waterfront, such as the Western Beaches Storage Tunnel, could be redistributed to a broader area to bring a potential reinvigoration along the whole watershed system.

The recognition and reconnection of the open spaces tracing the Garrison Creek ravine can also create a strong linkage between the neighbourhoods lining the network and the Toronto waterfront. This linkage can also create the potential for the waterfront to become more than a thin green line at the base of the city, extending itself instead, finger-like, into the northern neighbourhoods. The tracks of the ravines, Garrison and others, travel north into Toronto's inner and outer suburbs, and can become an important instrument, in partnership with public transit and 'main-street developments,' to provide a balance of open public space that supports community initiatives for new housing. As a newer expression of a co-evolving system, the intensification of main streets would be accompanied with a corresponding intensification and reconnection of many kinds of public open spaces. The increase of housing on Toronto's main streets requires an equal increase in public space amenities; adding to the network of open spaces that create linkage to the waterfront can provide that improved recreational breathing space.

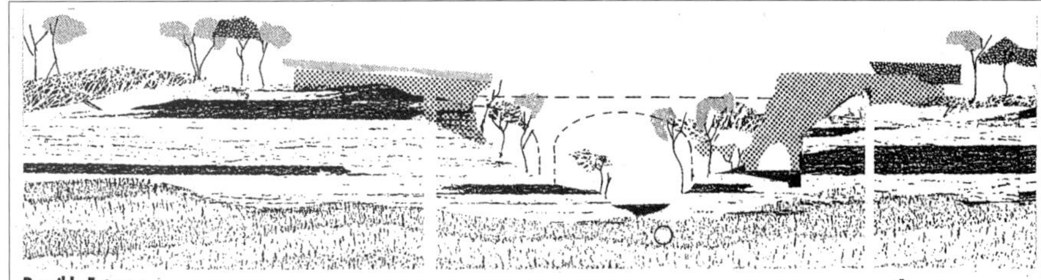

**Possible Futures:**

6

Growing community awareness of the Garrison Creek Ravine has started a movement towards the excavation of the original landform and revealing of the bridges where they occur on public park land. The considerable inventory of open space on the ravine route suggests the potential of a connected open space system that could knit the Garrison communities to Fort York at the original mouth of the creek and Lake Ontario. Illustrated here is a condition of the ravine and the city in a new balance, with new ponds recalling the original creek, excavation to the original ravine elevation, and the revealing of the connecting infrastructure of the bridge, that allows both the city and the ravine to co-exist.

The common thread in the scenarios we're proposing is that our 'secondary' water network of creeks and their ravines should become a critical component of how we think about our city. We should reveal buried networks in yet undiscovered urban public spaces, we should find new ways to renew ravines and waterways that have become degraded, and we should ensure that the preserved natural ravines that still exist can thrive through improved connections to neighbourhoods, streets and transit.

The reading of the buried Garrison Creek ravine is a reading of the city not as a nostalgic exercise, but as a new look at an ongoing dynamic partnership between the urban and natural landscape. Not the calm Victorian commemoration of how the city conquered the landscape through epitaphs, bronze letters and fish in the sidewalk, but an active transformation of the city, its public spaces, its landscape and its infrastructure – a new chapter in the co-evolving system of ravine and settlement and a new understanding of the dynamic role of water and infrastructure in the formation of Toronto.

{interview}

# The living machine: An interview with Helen Mills of the Lost Rivers Project

You might remember where you've lived by your street address. But Helen Mills remembers where she's lived watershed by watershed. In 1995, she and a small group of like-minded people started the Lost Rivers Project, a joint program of the Toronto Green Community and the Toronto Field Naturalists. Now in its thirteenth year, the group has named and mapped dozens of creeks and has accumulated an exhaustive web archive of information about these creeks. The group also leads the popular Lost Rivers walks, which take participants on storytelling tours of the city's buried creeks. In the past eleven years, 15,000 people have gone on over 400 walks. These walks tell the story of Toronto's past and present landscapes, people and neighbourhoods; connect people with information about urban water and ecosystems; and provide practical ideas about what citizens can do to help the environment. Herewith, an interview with Lost Rivers' Helen Mills: explorer, educator and environmentalist.

H<sub>T</sub>O: How did you become interested in Toronto's lost rivers?

HM: I started out, as a young person, being both very conscious and very unconscious of the physical fabric of the city, perhaps more so because I was an immigrant. I was conscious of the shapes and spaces of this place, which were different from all others I had known, but I was unconscious in that I accepted them as a given – 'just the way things are.' For example, I can remember being uncomfortably aware of the steep, seemingly man-made banks around the playing fields in Eglinton and Ramsden parks, but didn't question what had made them that way. I had no idea that there was something there that could be further interpreted and understood. My major mental map of the city would have been the street grid and the subway. Those were the important reference points.

I discovered that there was more to the story in the mid-1980s in a first-year U of T physical geography course. A lab on physiography and urban form precipitated my first recognition that the little anomalies in the grid that confused me long ago have a meaning and history that is decipherable and

*Helen Mills leads a walk along the Don on Canadian Rivers Day, 2008.*

discoverable. I was specifically blown away by the area round Queen, Sherbourne and Richmond streets. Jacob Spelt, a former geography prof at U of T, had written a paper called 'A Look at an Airphoto,' in which he talked about how the only evidence of the presence of Taddle Creek at that corner is a little perturbation in the path of Britain Street.[1] As a courier, I had driven my car, Hank the Tank, out of the gas station at Sherbourne and Richmond a few thousand times, each time giving myself whiplash as the clunky car moved down into a curving alley just across from Britain Street. I realized that the curve and dip were really relict bits of ravine edge!

All of a sudden my gestalt of the city shifted and I understood that those little bits of topography and strange variations in the grid are really significant. And that permanently changed my points of reference. I was horrified at what we had done to our rivers, despite the fact that I knew their burial had resulted in the end of deaths by cholera and other waterborne diseases. Suddenly, I wanted to make environmental art, to paint blue lines along the path of the creeks, to name them and bring them to the surface of people's awareness.

**So, at that time you found like-minded people – other geographers, people who were interested in water or just interested in making a positive environmental change in their neighbourhood?**

The idea actually sat on the back burner for about ten years, until 1995, when I discovered the Toronto Green Community (then the North Toronto Green Community) and found myself

1 J. Spelt, 'Downtown Toronto: A Look at an Air Photo,' *Canadian Geographer* (1966) vol. 10, pp. 184–89.

connected to hundreds of environmentalists working on water and other issues. In the group's first community meeting, I found myself in a subgroup whose members were talking about water, and Adele Freeman of the TRCA – best remembered by me for her role in organizing hundreds of volunteers to paint rivers on Yonge Street during Toronto's bicentenary celebration – handed me *Forty Steps to a New Don*. I read the vision statement about a beautiful river that the Task Force to Bring Back the Don was going to recreate, with sparkling clean water and birds and fish and people wandering under trees, and I just thought, 'These guys are definitely out of their minds. They are really crazy. And I want to work with them.' I never really looked back.

Which people or projects or ideas have really inspired your work?

One of the things that happened at the same time that I was doing my first-year geography course was that I walked through an exhibition in U of T's Faculty of Architecture of James Brown and Kim Storey's maps of Garrison Creek. This was another of those really profound epiphanies, I guess. I loved that imagery of the river superimposed back onto the street grid. I also saw a similar and very beautiful map by geologist A. P. Coleman in which he showed the shoreline of Lake Iroquois by colouring in the city grid with blue right up to the Casa Loma hill. These visual representations were very moving. And I've been very inspired by Coleman's writings: using very dense Victorian prose, he managed to convey something of the joy and romance he felt in exploring Toronto's distant past over a lifetime, and gave very specific geological and geographic information that enlivens every little interstice of the city. Did you know, for example, that the bones of a mammoth were uncovered on Dupont Street near Bathurst while a sewer was being dug there? Or that at Sherwood Park, when the village of North Toronto got its water from the springs in the hills there, the pumping station was driven by natural gas in the sandy hillside? These are just amazing little nuggets that change everything about how you see the city.

*A lost river: the upper reaches of Mud Creek.*

How has your understanding of water and water issues changed since you started the Lost Rivers project?

I think I have a better understanding now than ten years ago of the enormity and complexity of the issues. I revisited the word *ecosystem* when I first ran into it in David Crombie's *Regeneration* report and the TRCA's *Forty Steps to a New Don*, in which the 'ecosystem approach,' or the 'watershed approach,' to managing urban water were introduced. All the stuff that was written in that period had diagrams to illustrate the question 'What is an ecosystem?' And all the diagrams were a picture of a food web and a picture of a sun as the energy source driving the system. I kept looking at these things and thinking, 'What's this?' I knew that this approach was based on studies, including one I had read in the relatively distant past about the Hubbard Brook watershed in New Hampshire. The idea was that a watershed is a good unit to pick when studying bio-geochemical flows because it allows for the measurement of inputs and outputs in a defined area. So I thought, 'This makes really good sense.' But then I would see this food-web diagram and hit a mental roadblock.

So I looked up an old textbook – a book by Eugene Odum called *Fundamentals of Ecology* – with flow diagrams comparing the energy inputs and outputs of natural, agricultural and urban environments. This would be epiphany number I-don't-know-which. I realized, looking at them, that the key difference between an urban ecosystem and a natural ecosystem is fuel. As in the carbon that we're extracting from storage deep in the earth's crust and introducing into the earth's atmosphere at a rapid rate.

In an urban ecosystem, fuel has replaced the sun as the main source of energy that's driving the system, while in agricultural systems fuel is intermediate – the sun is still the major source of energy for photosynthesis – and in a natural system the sun is the primary source of energy, along with other natural energies like gravity and radiation from within the earth. Once I saw that 'other natural energy' piece of the diagram I was happy, because I had found the missing piece from those diagrams of urban ecosystems with the pictures of the sun and the food webs. I had the whole urban hydrological system down: gravity pulling the water down, the sun pulling it up through evaporation and evapo-transpiration, and fuel driving all the different ways that we've rearranged nature in the city. And it just beautifully made that connection to global warming: to this day, most of us don't realize the enormous energy cost of the water that we take for granted – that is our 'right.' Water uses up one-third of Toronto's energy bill and produces one-tenth of our greenhouse-gas emissions.

So, go figure. That's a lot of $CO_2$. That was my huge epiphany. You know, we got cholera. We figured cholera out. All these amazing engineers went away and sweated bullets and figured out how to make these extraordinary sewage systems and filtration plants and saved lives. Just so incredible, the work that was done – that was a huge paradigm shift, I think. I began to realize that we need another paradigm shift right now, having just about exhausted the potential of the present system, and having discovered some of the unintended consequences of our great city-building experiment. How can we now begin the shift to ecologically literate engineering that fundamentally alters the nature of our relationship with water? We can't keep using up this much fuel, putting this much $CO_2$ into the atmosphere just so we can have hot and cold running water and flushing toilets. Not to mention the contaminants that we've been introducing into the biosphere via air and water.

I almost hear you weighing which programs to invest in – stormwater management on one side, and water efficiency on the other. Do you come down on the side of water efficiency because of your energy concerns?

You know, I don't know if I would or not. What I'm trying to say is that both those systems belong in the category of a fossil-fuel–based, heroic engineering paradigm that probably has to, and is beginning to, shift dramatically. We need to look at completely different technologies. Examples might be the tap-to-toilet technology that was developed in Canada and is now in use in Singapore, or the new development in Holland where groups of fifteen houses share a small linear waterway that cleans all their sewage and recirculates it back as clean water. Ultimately the goal would be to live within the carrying capacity of local river systems, and to create an urban environment that mimics the functioning of a forested eastern North American ecosystem as closely as possible. So stormwater management and water efficiency are really taking care of different pieces of one ecosystem, and the goal should be to find ways to make better use of the free gifts of water that arrive via gravity, as well as finding ways to return the water we use to the water table close to where it fell, and cleaner than when it arrived.

I don't think it's just about daylighting creeks. We know the source of the problem is in the grid. So the question is, how do we rearrange it? It's a problem about the way the built environment is structured, similar to the problem of car-based sprawl. How do you rearrange everything about the structure of the city and the lives of its residents, given the enormous inertial mass of the existing structures and lifestyles, and the potential cost? I think those are the really interesting questions for the future. The challenge, as the Millennium Ecosystem Assessment concluded in its 2005 report, is to 'reverse the degradation of ecosystems while meeting increasing demands for their services as population grows.' It's an awesome challenge for the engineers and politicians who are in charge of the nuts and bolts of our everyday lives.

**Let's get back to the Lost Rivers project. When I go to www.lostrivers.ca, I notice that you're very focused on central Toronto. Is there any reason why you chose a certain space within the city to focus on?**

Because we started in North Toronto, and the idea was that we would understand our neighbourhood ecosystem. We started out looking at all the creeks that flowed through that space, which is a former bedroom commuter town that eventually merged with the city and became a planning district. Initially, we looked for the watersheds of rivers that once flowed through our local neighbourhood, on the basis that these would be the watersheds that North Toronto residents could directly affect. But everyone lives in a watershed, and as time went on, other people in other watersheds got interested and we developed fuzzy boundaries. If someone is interested in a creek and wants to hold a walk or research it, that's great.

**If you were walking down a street in search of a lost creek in, say, Malvern or Thistletown, what would you be looking for in the landscape? What are the clues?**

Look for a shopping mall. Lou Wise, a pilot who used to work very closely with Charles Sauriol, once phoned me up to tell me that he was flying the Rouge watershed for Adele Freeman and taking air photos as he went. He told me that he kept flying up the Rouge's tributaries and then, poof, the river would disappear and there would be a shopping mall.

Ultimately the best way to find a creek is to look for topography. Water is all about gravity – Odum's 'other natural energy.' If you see a whole lot of hills and valleys, ask yourself, 'Where's the water?' Even in Toronto, which is legendary for not having much topography, there are telltale ups and downs all over the place.

**Okay, let's turn back to the local. We understand that Mud Creek is one of your favourite lost rivers. What makes it special?**

I think it's special because of the Don Valley Brick Works, and because it traverses some interesting human and natural history. Above the 401, Mud Creek lies in a landscape of muted till-plain tablelands created by the action of the great ice sheets,

*A found river: Mud Creek at the Brick Works.*

a home to remnants of primordial maple-beech forest in Earl Bales Park. This area is a meditation on the impact of twentieth-century fuel-based transportation systems – trains, planes, subways and expressways – a no person's land designed on the scale of the car that necessitated the wholesale destruction of forest, wetlands and 10,000-year-old soils. It bears the scars of fifty years of occupation by the armed forces and is home to many individual tales of war and peace, the stories of World War II survivors, many of them from concentration camps, who came here to make new lives.

Below the 401 lies the urban history of Toronto in the first part of the twentieth century, a tale of ever-expanding sprawl, of the transition from Garden City to ticky-tacky suburb, of expansion beyond the carrying capacity of successively bigger watershed areas. Further south, there are sites that tell of earlier human settlements, of the miraculous parallel evolution of agricultural systems in the New and Old worlds – I'm thinking here of the Quandat archaeological site near Eglinton Park – of fur traders, ancient trails and the migrations that took place during the French era. It's the story of British colonialism, of the clearing of the forests, the cadastral survey, of a time when Ontario was the breadbasket of Canada, of the coming of the gentlemen farmers. Near the

Brick Works, Ernest Thompson Seton camped in the woods and wrote about walking up and down the ravine. At the Brick Works, Coleman worked out the events of the past million years in this part of the world. The Mud Creek watershed is a study of the ways in which machines, not wind or water, are now the main factor forming the landscape, and of the many ways in which we have distorted the flows of water and materials in cities.

So, looking toward the future, what would Mud Creek look like in a fully restored state?

Are we on to fairy tales now?

Well, you know, you were captured by *Forty Steps to a New Don,* which pushed you toward action. How would you inspire people to rethink Mud Creek – all the tiny tributaries, rather than just the big trunks of the river system?

Okay, well, here is one small fairy tale: Downsview Airport is at the headwaters of many small creeks that flow east into the Don or west into Black Creek. Mud Creek rises between two former runways. Yellow Creek starts between another two sets of runways, Burke Brook between another two sets of runways, and so on. I imagine Downsview Park as being something unique anywhere in the world: an urban national park, with some real meaning and clout behind those words. A national park that is a laboratory for a future sustainable urban environment, a museum of the twentieth-century urban environment both good and bad, a place that tells the stories of local communities, their links with twentieth-century history, World War II and the post-war era, and also the older stories of First Nations and French communities now lost.

I like the concept of the winning design for Downsview several years ago – Tree City[2] – because it envisaged links into the urban forest beyond the park boundaries. I imagine a restored fabric of linear linkages along the line of all the creeks, so that when you're in Downsview Park, you're in the national park. When you're anywhere along the course of Mud Creek, you're in the national park. When you're in the Brick Works, you're in the national park.

2 www.pdp.ca/en/corporate/concept.cfm.

I imagine signage and linkages, walking trails, migrating birds and butterflies. I would love to see a trail along Mud Creek from Downsview Airport all the way down to the Brick Works. Perhaps it could be called the Coleman trail because Coleman made his groundbreaking geological discoveries at the Brick Works and is buried near Mud Creek in Mount Pleasant Cemetery. I would love even more to see the development of a community based on restoration of the watershed, perhaps a version of the Dutch linear-water cleaning system installed all the way down Mud Creek that uses rainwater where it falls, cleaning the water every fifteen houses and returning it to nature cleaner than it was when it arrived in the neighbourhood.

I imagine the creation of a new kind of life-support matrix for the city based on natural systems and energy, not on inputs of fossil fuels. Recently, financier Kenneth de la Barre, who spoke at Confluences, a conference held at the University of British Columbia in November 2007, commented on the possible cost of daylighting streams or other similar efforts by pointing out that green infrastructure is exactly where we should be putting our dollars: 'It's only money and it's what we should be spending money on.'

Thanks to Peter Hare, Ed Freeman, John Wilson, Richard Anderson, Ian Wheal, Sandy Cappell, Daniel Spring, Georgia Ydreos, Liz Forsberg, Jessica Laplante, Dagmar Baur, Emily Alfred, Justyna Braithwaite, Katie Harper, Natalia Crowe, Ellen Kessler, Michelle McMahon, Dick Watts, Madeleine McDowell, Nick Eyles, Bill Snodgrass, Michael Gauthier, John Rouse, Christine Tu, Peter Heinz, Eduardo Sousa, Michael McMahon, Margaret Buchinger, Mark Wilson, Adele Freeman and many others, as well as the following organizations: the Toronto Green Community, Toronto Field Naturalists, Evergreen, Roots and Shoots, LEAF, RiverSides, Human River, the Don Watershed Regeneration Council, the TRCA, the Waterfront Regeneration Trust, Brown and Storey Architects, the Community History Project and the Taylor Massey Project.

John Lorinc

# The big gulp: How Toronto's Wet Weather Flow Management Master Plan (a name no one likes) will save the lake

On a balmy spring day, Janice Palmer strolls down a winding ravine road toward Moccasin Trail Park, a secluded saddle of open space near Don Mills Road and Eglinton Avenue where the City has put nature and engineering to work to slow stormwater as it drains off a nearby townhouse complex.

Palmer, a retired high school teacher, is one of the two vice-chairs of the Task Force to Bring Back the Don. Along with hundreds of volunteers, she's worked for years to re-naturalize Toronto's ravines and make them more 'absorbent.' The Task Force and other groups have reintroduced native species and built wetlands to reduce the amount of polluted runoff that flows through storm sewers and into the Don's tributaries. These days, they're working on even more ambitious schemes, such as the recently built stormwater-management project in Moccasin Trail Park.

This undertaking doesn't look like one of the Task Force's bucolic regenerated wetlands, although it plays a very similar ecological role. The pipe-shaped park follows a steep gully that runs down the west bank of the Don Valley. At the bottom, it opens onto an old stand of apples trees, a bit of grassland and a trail that leads under the highway. Until 2002, a stand of scruffy trees formed a buffer between the little-used park and the Don Valley Parkway.

Today some of those trees are gone, and in their place sits a banana-shaped pond that's fed by a creek that runs down the ravine wall. This engineered body of water is divided into sections separated by submerged berms. As stormwater flows into the pond, sediment and suspended pollutants sink to the bottom, and the cleansed water flows over one berm into the next section before draining into a culvert leading to the Don River. Along the creek and up into the ravine, the City has installed piles of rock and swales to slow the water. The principle, says Palmer, is to 'keep the sediment back so it doesn't end up in the Don.'

Palmer and other Bring Back the Don volunteers were active partners in planning these improvements to Moccasin Trail Park, but the work was all done by the City's water department. What's more, the City of Toronto financed the project using

*Storm over west Toronto, 2008.*

Section 37 grants from a nearby development on the edge of the ravine (the planning department often asks builders to make financial contributions to community projects in exchange for additional density or height). 'This wouldn't have been constructed had it not been for the increased hard surface up on the tablelands,' says Palmer as she strolls along the banks of the pond, surveying the changes with satisfaction.

Palmer describes the Moccasin Trail pond as 'a small piece in a very big puzzle' – a puzzle that comes with an irredeemably bureaucratic moniker: the Wet Weather Flow Management Master Plan. In spite of the ungainly title, the WWFMMP (hereafter referred to as the Plan, but also now known around City Hall as 'the Water Pollution Solution') is easily Toronto's most far-reaching strategy for managing rainwater and snowmelt.

Yet the Plan is anything but a traditional works scheme. Approved in 2003 after years of planning and public consultation, the twenty-five-year, $1 billion strategy will radically alter the city's relationship with its water resources through a pragmatic marriage of environmental activism and twenty-first-century civil engineering. It encompasses everything from downspout disconnect and basement flooding remediation programs to dozens of wetland and stream-restoration projects,

*Stormwater Performance on Various Ground Surfaces*

| | Stormwater absorbed into the ground % | Runoff % | Evaporation/ transpiration % |
|---|---|---|---|
| Natural groundcover | 50 | 10 | 40 |
| 10-20% impervious (exurban) | 42 | 20 | 38 |
| 35-50% impervious (suburban) | 35 | 30 | 35 |
| 75-100% impervious (downtown) | 15 | 55 | 30 |

Source: William Snodgrass, *Intake Protection Zones (IPZs) in the Great Lakes: Scientific, Environmental Engineering, & Practical Considerations for Protecting Sources of Drinking Water*. Presentation, Source Water Protection Symposium, April 2007. www.trentu.ca/iws/documents/SWP_Snodgrass.pdf

as well as the construction of major new trunks to prevent massive discharges of sewage into the lake during heavy storms.

The Plan's guiding principle is to restore natural hydrological cycles within the context of a highly urbanized environment. Throughout the long implementation phase, City officials will stress the use of environmentally sustainable means to slow or absorb stormwater runoff rather than engineered end-of-pipe solutions. The adoption of the Plan was a watershed moment, so to speak. As Councillor Glenn De Baeremaeker, an environmentalist who, at publication, chaired the Works Committee, says, the strategy is nothing less than 'a profound and fundamental change in how the city sees the world.'

Shortly after the 1998 amalgamation of the City of Toronto, works officials held a briefing for new councillors to introduce them to the water issues facing the nascent megacity. Former Etobicoke councillor Irene Jones recalls being both overwhelmed and 'aghast' by what she learned: not only was the amalgamated city's wastewater infrastructure enormous – 5,200 kilometres of storm sewers, 4,440 kilometres of sanitary sewers and 2,600 stormwater discharge points (a.k.a. outfalls) – it was also crumbling after years of financial neglect. 'There was a recognition that if they didn't do something to protect it,' says Jones, 'the infrastructure would collapse and the drinking water would be at risk.' As Jones points out, residents of the former local municipalities of Etobicoke, York, East York, Scarborough, North York and Toronto paid water bills designed to cover only the cost of operating their systems, which meant that long-term capital needs were not being addressed.

Lack of investment wasn't the only problem. Enormous growth in the 905 municipalities, especially Peel and York regions, had placed tremendous pressure on Toronto's watersheds, which extend well beyond the 416 borders. Moreover, the region had rapidly lost its tree cover, thus foregoing a critically important source of stormwater absorption. As De Baeremaker notes, about 17 percent of Toronto has forest cover, while in Markham, only 3 percent of the canopy remains.

Consequently, the water-borne detritus of suburban expansion – pesticide residue, road salt, rubber, oil, gas, paint, phosphates and whatever toxic crap people poured into the catch basins at the curb – was draining straight into the rivers, and thus the lake. Moreover, this water was tainted by high levels of E. coli bacteria from fecal material due to the combined sewer network. (In the old City of Toronto, there are 1,290 kilometres of combined pipes, which contain both stormwater runoff as well as household sewage.) One analysis found the E. coli density in combined pipes was 1.2 million parts per 100 mL, and 400,000 parts per 100 mL in stormwater pipes. In comparison, treated wastewater from the Ashbridges Bay plant has 1,000 parts per 100 mL, while the standard for waters used in non-contact water recreation sports such as canoeing is 500. The city's beaches are closed when the count exceeds 100.

At the instigation of the works department, City Council established a working group of about forty engineers, hydrologists, consultants, environmentalists and citizen representatives to develop a 'wet weather flow master plan.' De Baeremaeker says the unwieldy name of the undertaking bespeaks its bureaucratic origins. For two years, the working group made little progress as they grappled with the mammoth task. Some participants wanted to dig up and separate all the combined sewers. Others were steadfastly opposed to engineering solutions. By the time Irene Jones stepped in as chair in 2001, the meetings had become mired in argument. Nothing was happening: as she recalls, they couldn't even agree on the minutes of the previous meeting.

A left-leaning councillor representing a lakefront Etobicoke ward, Jones whipped the committee into shape. She dismissed

BIOSWALES AND BLACKTOP

Traditionally, parking lots are contoured around slight depressions that convey runoff (a brew that includes salt, sand, oil and gas, and rubber residue) into catch basins that are tied to the local storm sewers. Bioswales offer a much more sustainable alternative. The blacktop is contoured so runoff flows to gently sloping engineered ditches built at the edges of the parking surface. These are filled with gravel and topped off with organic material, including hardy plants. The runoff flows into the swales, which are designed to filter suspended solids and other contaminants before returning the water to the soil and/or the storm sewers.

unrealistic demands (installing separate storm and sanitary sewers throughout the old city, Jones told the group, would be enormously costly and could literally take 200 years to complete) and gave voice to activists from the old City of Toronto who'd been involved with the Bring Back the Don movement. Lastly, on Mel Lastman's right-leaning council, she took care not to speak out about the project for fear of attracting opposition from her conservative colleagues.

As she steered her group toward a final report to Council, an alarming proposal surfaced from the right: a plan to privatize the City's water operations. In the aftermath of the Walkerton water scandal, Jones knew that such a change was politically explosive. But when she asked around, she discovered the City's water officials wanted their department to be reorganized as a self-sustaining business unit with more control over its revenues and expenditures. As it turned out, the question of ownership was neither here nor there. Jones persuaded the City's finance staff to consolidate all the water-related budget expenditures. It would prove to be a crucial step in the eventual implementation of the Plan because, for the first time, City officials could calculate how much additional revenue they'd need to raise in order to finance improvements.

In 2003, during the waning days of the Lastman regime, Council finally approved the Wet Weather Flow Management Master Plan. Jones, who had credibility on both the left and the right, succeeded in building a broad-based consensus on Council. As Katrina Miller of the Toronto Environmental Alliance notes, 'It was all about political will.'

There's little doubt the Plan was one of Council's stellar achievements in the post-amalgamation period. Yet five years on, very few Torontonians know of its existence, much less understand the details and long-term implications of the entire strategy.

The Plan has thirteen strategic objectives. These include meeting water and sediment quality guidelines; virtually eliminating toxins through pollution prevention; improving aesthetics and promoting beach swimming; reducing erosion; re-establishing natural hydrological cycles and minimizing flood risks; protecting habitats and reducing fish contamination; and eliminating the discharge of untreated sanitary sewage and reducing basement flooding. The strategy also calls

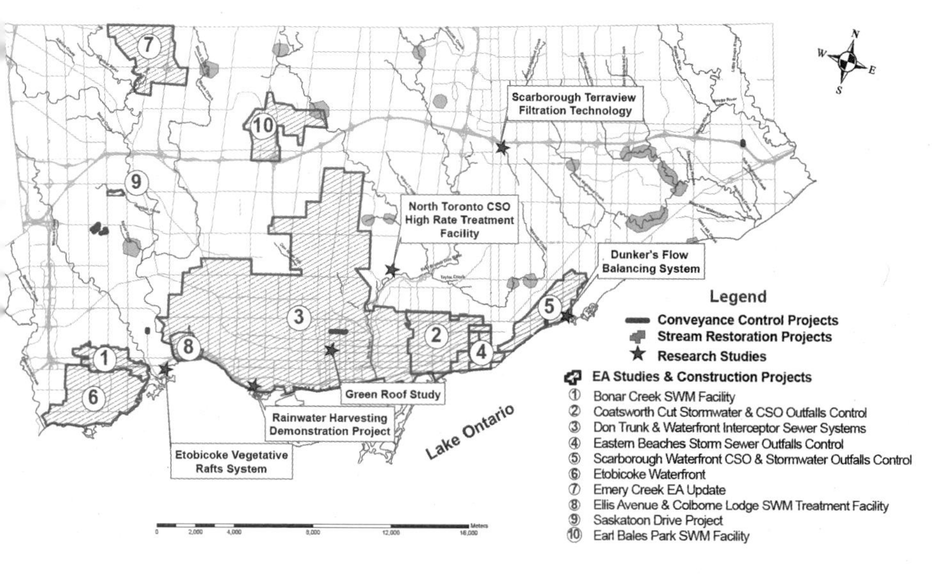


*Wet Weather Flow Management Master Plan project locations, 2007.*

for extensive public education – all the pithy ad campaigns in recent years, such as those featuring a fish admonishing homeowners to stop dumping chemicals into sewers, flow from the Plan's push to create more understanding about such connections. Taken together, it's an unlikely bureaucratic stew of can-do engineering, ecological activism and feel-good civic propaganda.

In practice, the Plan's particular goals include extensive basement-flood protection retrofits, stream and aquatic habitat restoration and a bevy of major civil engineering projects to improve the pipes. Council approves the water department's annual work plan, which derives from these long-term targets. According to Michael D'Andrea, director of water infrastructure management for Toronto Water and the official in charge of implementing the Plan, these far-ranging projects are being phased in over five-year periods, beginning with downspout disconnects, the construction of new wetlands, bioswales and stormwater-management ponds (such as the one in Moccasin Trail Park) and the restoration of streams that have suffered extensive erosion. The latter involves reforestation, bank regeneration and the removal of fish barriers. The City hires contractors, landscapers, hydrologists and environmental engineering firms to do the actual work. Interestingly, City officials spent the first couple of years looking for specialized

*Storm damage to Finch Avenue West, 2005.*

consulting firms to bid on these gigs, which are somewhat atypical by the standard of conventional municipal waterworks contracts. As of 2008, however, the City has lined up contractors who can do the work and has initiated ecological projects such as re-naturalizing stream banks.

The pipe improvements outlined in the Plan are more conventional. The City wants to build new storm sewers in the Don Valley and intends to physically disconnect existing storm and sanitary pipes by removing connections leading from one system to the other. The Plan also calls for the installation of deliberately 'leaky' storm pipes that will allow some runoff to filter back into the ground beneath city streets, where the soil can filter out impurities naturally.

This approach contrasts with earlier engineering-driven attempts to clean up the waterfront. In the 1990s, the City embarked on the construction of a handful of very large holding tanks near the eastern and western beaches designed to capture excess runoff before it reached the lake. The one near Sunnyside cost $75 million and was completed in 2006. Two others, near Ashbridge's Bay and Balmy Beach, can hold enough water to fill six and twenty-one Olympic-sized pools, respectively. The tanks allow suspended solids to settle out before the water is released. These highly engineered end-of-pipe solutions have improved beach-water quality, but they still don't deal with the stormwater at source, nor do they provide natural filtration.

FACING PAGE
*Black Creek meets Finch Avenue West, August 19, 2005.*

In many ways, the strategy was ahead of its time, and has meshed well with the environmentally conscious politics of recent years. David Miller inherited the Plan when he became mayor in early 2004. The new Council, more left-leaning than its predecessor, has placed a heavy emphasis on making the City of Toronto a green municipality. The Millerites immediately recognized the opportunity presented by Irene Jones' ideas for dealing with stormwater. Early in the first term, they put their money where their political mouths were, approving a capital investment surcharge on homeowner water bills with 9 percent annual increases through 2017. The revenues will finance the implementation of the Plan. As De Baeremaeker notes, the City's annual water budget in 1999 was $149 million; by 2017, it will have climbed to $900 million. 'That's how visionary this document is,' he says.

On August 19, 2005, a torrential summer storm roared across southern Ontario, dumping fifteen centimetres of rain in a three-hour period. The downpour caused $500 million in damage – one of the worst natural disasters recorded in this part of Canada. 'It was the single biggest storm we've had since Hurricane Hazel,' observes D'Andrea, a genial and poised engineer who is acutely aware of the legacy of Hazel and water catastrophes in other urban areas. A chunk of Finch Avenue West between Jane and Keele was completely swept away. In ravines across the northern part of the city, hugely swollen rivers devoured their banks. The rush of water was so intense that it destroyed a sanitary sewer buried beneath the bed of East Highland Creek, causing untreated human waste to be released directly into the surging stream.

Through much of North York, Etobicoke and Scarborough, sewage backed up into more than 3,600 basements. A less severe storm in May 2000 had produced a similar result. 'The public is irate, and rightly so if you've had sewage in your basement repeatedly over a number of years,' says D'Andrea. 'The system wasn't designed for these extreme events.'

Climatologists talk about storms in terms of probabilities – the likelihood that a storm of a given intensity will occur over a defined period. One of the consequences of global warming is that high-intensity storms, once considered relatively rare, now occur more frequently. As D'Andrea parses the problem, Toronto's stormwater-management infrastructure is engineered to handle runoff volumes generated by the sort of storm that comes along once every two to five years. The August 19 event, however, was considered to be at least a '100-year storm.' And the fact is that 100-year storms are no longer as rare as they once were. Indeed, when water officials checked the history books, they realized that since the mid-1980s, Toronto had endured at least eight storms that exceeded the City's design standards.

As Hazel did half a century ago, the August 2005 storm served to refocus municipal attention on the need to find better ways to manage runoff and mitigate risk. Unlike in Hazel's time, however, the City of Toronto had already been thinking hard about these very challenges. That the Plan existed on August 19, 2005, is a testament to the critical importance of forward planning in the era of climate change. But there's little doubt the storm has altered the way the Plan is being implemented.

When D'Andrea's staff analyzed the causes of all the basement flooding, they realized the homes affected were situated in low-lying areas north of the 401. As he says, that part of the city is a 'series of soup bowls.' What happened was that due to the volume of rainfall, storm sewers and street catch basins overflowed, causing the runoff to pool in low-lying areas. Water also poured into the manhole covers that provide access to sanitary sewers, thus inundating the pipes that carry human waste. What's more, hundreds of these flooded homes had similar symptoms of vulnerability: no back-flow valves in basement drains, cracked foundations, poorly graded yards, reverse-slope driveways leading to sub-grade garages and connected downspouts.

In the years since, water officials have focused intensively on retrofitting those parts of the city against future flooding – it's a political issue as much as anything else. D'Andrea's department is spending $16 million to deal with basement flooding between 2006 and 2010. There are new programs to help homeowners retrofit their homes and disconnect downspouts, and a total of 26,000 homes have been uncoupled since 2003. The City has moved to improve street sweeping to prevent catch basins from becoming clogged with leaves. And there are plans afoot to construct larger-volume storm sewers and underground storage tanks as well as to create 'dry ponds' in local parks where excess runoff can be directed during heavy storms. 'All of these infrastructure works are costly,' says D'Andrea, 'and they're

*Ellis Avenue stormwater pond, with the Humber River and Lake Ontario in the background, 2008.*

only put to use in extreme storms.' But, as he points out, there's a 'huge premium' for repairing the damage caused when proper protections aren't put in place.

Although the City is hustling to deal with the basements, water officials haven't forgotten about the other pieces of the Plan. Stormwater-management ponds and wetlands are being constructed on the Etobicoke waterfront, in North York's Earl Bales Park and in High Park. Creeks that were ravaged in August 2005 are being rebuilt with natural channel designs, bank revegetation, wetland restoration and the removal of fish barriers. (East Highland, which required extensive remediation, is due to be finished in 2009.) There's a new marsh at the mouth of the Humber River. Meanwhile, the Plan's budget also allows community-level environmental groups to secure grants up to $25,000 to help them complete small-scale water-improvement projects – a strategy Jones insisted on as a way of increasing public engagement with municipal ecological initiatives; local groups have helped with the restoration of East Highland Creek, for example. Lastly, city-planning officials are pressing ahead with green building standards, approved in January 2007, for new construction that will require developers to find ways to do things like recycle rainwater and use drought-resistant plants in landscaping plans. It's all of a piece.

By far the biggest project related to the Plan, however, is the proposed Don and Waterfront Interceptor Trunk, a classic large-scale civil engineering project. In the 1950s, Metro built what was called the Coxwell Sanitary Trunk deep beneath the

TOP *Where stormwater enters the system: just one of 122,500 catch basins, 2008.*

ABOVE *Where stormwater leaves the system: Woodbine Beach, 2007.*

Don Valley. Sanitary and combined sewers from about a third of the city, representing 750,000 residents, are directed into this massive pipe, which conveys 400 million litres of untreated wastewater to the Ashbridges Bay plant each day. As the city prepares to increase its population by another million people over the next quarter-century, the Coxwell trunk's capacity won't be sufficient to handle the growth. And with so much medium- and high-density mixed-use development slated for the waterfront and the Port Lands, D'Andrea says the City must take steps to prevent huge discharges of untreated sewage into the Don when heavy rains overwhelm those old combined sewers.

The plan is to twin the interceptor with twenty-five kilometres of new trunk pipe constructed beneath the Don Valley and along the waterfront; a major environmental assessment to do this was underway in 2008. D'Andrea says the trunk twinning, the most comprehensive works project in half a century, could cost $500 million or more. The City doesn't yet know where that kind of green is coming from, but D'Andrea says the strategy is to persuade the federal and provincial governments to contribute, using funds earmarked for Great Lakes preservation and strategic infrastructure.

No large policy program remains static, especially one with so many moving parts. But it's a testament to the new thinking within the City bureaucracy that the broad political consensus behind the Water Pollution Solution has endured, even solidified, in the years since it was first approved. Environmentalists like Katrina Miller and naturalists like Janice Palmer don't balk at engineering solutions, and the engineers working for Michael D'Andrea have found religion with regard to incorporating natural systems and ecological planning into their thinking about municipal works. Indeed, the big 2005 storm and all the attention being paid to climate change has heightened public and bureaucratic interest in these new approaches. 'Now, it's almost like the politicians are running ahead of the [Plan], and they're pushing it even further than it was ever intended to go,' says Toronto Environmental Alliance's Katrina Miller.

By way of example, Miller cites policy proposals by the City to impose green building codes designed to mitigate stormwater runoff – a relatively new addition to the Plan. And she also mentions how the water department has agreed (at the behest of Mayor Miller) to indefinitely defer another element of the Plan – the construction of a long 'deflector' at the mouth of the Humber River. Aerial photos show a long arching plume of murky water that flows out of the river and drifts toward the Western Beaches and the central waterfront after big storms. The deflector is essentially a breakwater that would extend far out into the lake from the spot where the east bank of the Humber meets Lake Ontario. Environmentalists hated the idea and lobbied against it. As Miller says, it wouldn't reduce water pollution, but rather spew it further into the lake in the way that higher smokestacks only dissipate emissions. 'Sunnyside Beach would be open more often,' she says, 'but Lake Ontario water quality would continue to degrade.'

Beyond the back-and-forth about the deflector, the effectiveness of the Plan remains constrained by the fact that the rivers running into the lake extend well beyond the borders of the City of Toronto; many of the problems begin upstream, beyond the reach of the Plan's innovative mix of solutions. The TRCA, which manages watersheds across the GTA, is well aware of the problem and does what it can to improve stormwater management in the 905 region. But the reality, as D'Andrea and Miller know, is that the detritus all ends up flowing into Toronto's lakefront.

The official line from City Hall is that we should be doing as much as possible to mitigate what happens south of Steeles, always with an eye to rehabilitating Toronto's beaches. We've all seen those iconic 1920s photos of swimmers swarming Sunnyside and we've all heard the tales of how children could ride for free on streetcars bound for Toronto's Riviera. Eighty years on, the City is promoting the Plan with the promise of swimmable beaches. (Never mind that the water off Sunnyside Beach was probably quite filthy back in the day.) But some who know the Plan from the inside out see this particular commitment as little more than a sales pitch – a nostalgic and easily understood symbol meant to front for an enormously complex undertaking.

As Irene Jones claims, the point of the Plan was never about swimming. 'It's about safe drinking water.' And therefore it is about nothing less than Toronto's future.

Eduardo Sousa

# Re-inhabiting Taddle Creek

*O, mighty Taddle! I stand for the last time at thy side, and look, as far as thy manifold indescribable impurities will allow, into thy dark depths. Strange thoughts come and go in my disturbed mind – tears start to my eyes; and something – I know not what, affects even my nostrils . . .*

*No more shalt thou behold the varied scenes upon thy banks, nor thy mighty influence. Nor more shall the sight of thee inspire noble thoughts. But thy work has been accomplished. Thou goest down to thy grave, unknowing and unknown. And the spectre Typhoid, the demon of thy banks, is correspondingly disappointed.*

'A Graduate's Farewell,' *The Varsity*, 1883

In her book *In Service of the Wild*, bioregionalist Stephanie Mills defines re-inhabitation as 'learning the whole history of one's bioregion or watershed, and developing a vision of sustainable ecological community from that knowledge, and from what we have been learning, in the last half-century, about elegant techniques of constructive gardening, recycling, energy conservation and waste treatment; ways of sophisticating old-style household and neighbourly frugality.' Furthermore, she says, re-inhabitation involves undertaking activities at the community level that enrich 'the life of that place,' through restoration efforts and other projects that connect people to place.[1]

Some years ago, I worked with a number of community groups whose activities aimed to reconnect people to a place that once functioned as a watershed. What follows is a recounting of the history of the Taddle from the beginnings of European occupation to its present incarnation as a sewershed, and suggestions for how we might re-inhabit the creek's watershed and the greater watershed around us to avoid the further degradation of these habitats and encourage the restoration and regeneration of damaged ecosystems.

1 Stephanie Mills, *In Service of the Wild* (Boston: Beacon Press, 1995), p. 7. In the second quote, she cites activist Peter Berg and ecologist Raymond Dasmann.

*McCaul's Pond, c. 1876.*

The history and mythology of Taddle Creek are entirely intertwined with the history of the old Town of York and, thus, with the City of Toronto. However, though it played a major role in the formation of the original settlement, the name Taddle Creek does not consistently show up in maps or estate plans of Toronto because the creek had different names along its route. For example, although it has been called Taddle Creek in Wychwood Park and the Annex for the past century, it has also been elsewhere referred to as University Creek, Brewery Creek and Town Creek, and as a brook on some estate plans.

Over 150 years ago, the Taddle flowed through two forms of subdivision and development taking place simultaneously on the land that was to become Toronto. The first were the large 100- and 200-acre agricultural lots and country estates from Davenport Road south to Queen Street that were given to the elite and governing classes of Upper Canada. The

A TADDLE TALE: McCaul's Pond

In those simpler days it was not unknown for undergraduates to spend spare moments beside the pond picking wildflowers and chasing butterflies. Some caught chub and shiners and the occasional speckled trout in its water. In winter the pond made a natural skating rink and the slopes beside were popular for tobogganing. In spring young lovers found it a romantic rendezvous, and in summer families watched while youngsters sailed toy boats on its surface. At least one student prankster made use of it to hide the College lawn mower under several feet of water, where it remained until the pond was drained years later.

Ian Montagnes, 'Taddle Tale,' *The Graduate*, Sept./Oct. 1979 second was the original Town of York as it was first laid out in a grid of ten blocks enclosed by the present Front, George, Duke and Berkeley streets just north of Toronto Bay.

Residents and breweries used the Taddle as a source of fresh water but, by the 1830s, there were already references to the foul smells resulting from the dumping of human and agricultural wastes. That said, the Taddle did not begin to fully disappear until the 1850s – maps indicate it was still emptying into Toronto Bay as late as 1855, albeit in a channelized form. In essence, Taddle Creek was a 'town and country' creek, and as the country developed into a town, the creek disappeared.

The exact location of the headwaters of Taddle Creek remains open to speculation. Some local historians believe the headwaters lie just north of the Lake Iroquois shoreline in Wychwood Park, where a pond and outflowing stream, locally referred to as Taddle Creek, are found amidst the remnant oak and white pine forest [1] (map). The pond is fed both by underground springs and by a stream located just to the north of it, most of which has been piped for over eighty years. In fact, the pond was created in the late 1800s by damming Taddle Creek and enlarging its natural basin, a project that was undertaken for aesthetic reasons and to supply ice for nearby residential iceboxes.

As it flowed out of Wychwood Park, the creek crossed what became Davenport Road to head due south through land that would eventually become a market garden belonging to a pioneering family, the MacNamaras, and which today is the site of the TTC's Hillcrest Yard. From there, the Taddle cut a southeasterly path to Lake Ontario through a landscape that had originally been a heavily forested, rolling lacustrine plain left behind by Lake Iroquois, to empty into Lake Ontario between Parliament and Berkeley streets, south of Front Street.

However, other than a couple of maps, very little evidence exists that there was a direct link between the Taddle in Wychwood Park and the Taddle that

A TADDLE TALE: Taddle Creek and food production

'John [MacNamara] grew at least cabbages, corn, carrots, squash, rhubarb, raspberries, apples and asparagus, with much assistance from his large family and several employees. The children were expected to weed a row of carrots on the way to school, and another on the way back. The produce was then taken by wagon along Davenport Road to Yorkville, or south on Bathurst Street to the Toronto markets.'

Jane E. MacNamara, *One Toronto Irish Family: The MacNamaras*

[1]

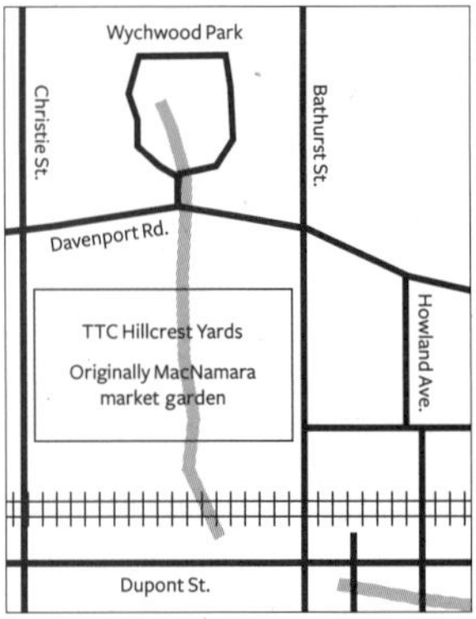

shows up in maps in the Annex. The lack of historical evidence for the creek's path may lie in the physiographic conditions of the land: the mixture of clay and sandy soils and the flatness of the area would not have provided good drainage for the creek. It is quite possible that Taddle Creek did not have a defined path through this area. Amanda McConnell, a long-time resident of the area who has absorbed much Taddle lore, has said the area was once a huge marsh without any predefined creeks running through it. And, though it's possible to pick up the Taddle in early maps slightly south of Dupont Street, between Albany and Howland avenues, there are, for the most part, very few recorded references to the Taddle flowing above ground through the Annex, just records of flooded basements and backyards and small bridges here and there.

A TADDLE TALE: Bogged down

'Workmen who dug an immense hole at Huron and Lowther for an apartment building ran into a bog – probably left by the Taddle – and went no further. The City bought the land, filled it in – and the Annex wound up with an unexpected park.'

Rae Corelli, *The Toronto That Used to Be* (Toronto: Toronto Star, 1964)

Picking up the Taddle at Howland Avenue [2], we trace its path east toward Kendal Avenue, slightly north of Lowther Avenue (the right-angle turn Kendal makes there was to accommodate the Taddle). Passing Spadina Road, it eventually cut directly across the Huron Street playground, a local parkette originally too marshy to build on. It continued to meander east along Lowther, cutting through what became yet another City park, Taddle Creek Park, then crossed Bedford Road and flowed south to Prince Arthur Avenue to feed a pond located in the backyard of the Women's Art Association before traversing Bloor Street to head into the University of Toronto lands. Today, that pond is drained, and the backyard is an asphalt parking lot.

[2]

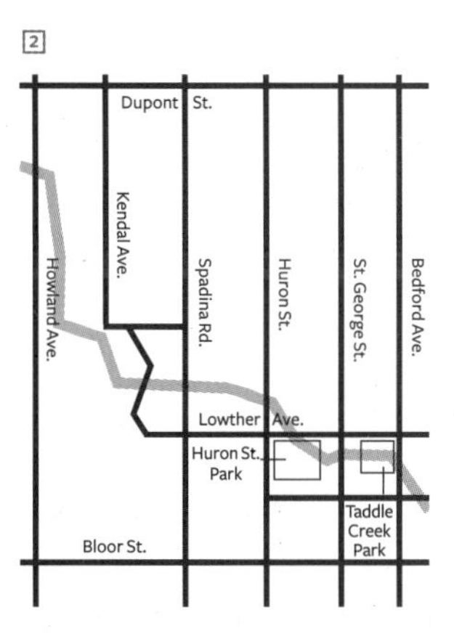


A number of commercial operations in this area have had problems with groundwater, including the store just south of this former pond, Remenyi's House of Music. Local lore has it that they struggled with flooding in their basement for years. The Taddle makes itself known on the Bloor subway line too; many sump pumps and other engineered solutions have been installed to deal with the Taddle there.

A TADDLE TALE: You can't keep a good creek down

'[Taddle Creek] caused enough havoc in the Annex. Without warning, it would burst through a cellar wall and flood it to a depth of three feet. Then it would recede only to reappear months later several blocks away as the culprit in a sudden street cave-in.'

Rae Corelli, *The Toronto That Used to Be* (Toronto: Toronto Star, 1964)

On Bloor Street, at the formal entrance to Philosopher's Walk, between the Royal Ontario

3

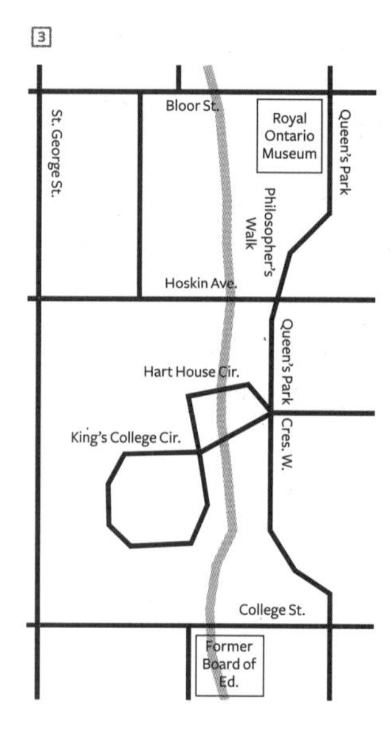


A TADDLE TALE: '... to disappear from the face of the earth ...'

'The City Commissioner has reported to the Council that a sewer is imperatively needed in connection with our meandering Taddle, or if no sewer be built then that the Yorkville people should be prevented from fouling the stream. The subject has often been before the Council, and we are glad to see that the Commissioner has viewed the stream from a sanitary, and not from a romantic, standpoint. He thinks the Taddle should be made to disappear from the face of the earth.'

*The Varsity*, December 9, 1881

'The attention of university authorities should be drawn to the state of the pond near the entrance to the grounds. The evil odour arising is simply intolerable. At this time of the year, fevers and sickness are so prevalent, too much care can hardly be given to drains, etc. Perhaps the newly appointed Medical Health Officer, Dr. Canniff, might do something in the matter. But certain it is that something should be done, and done speedily.'

*The Varsity*, March 17, 1883

Museum's Michael Lee-Chin Crystal and the Royal Conservatory of Music, the remnant ravine topography and pathway render visible the former path of the Taddle 3.

If the Taddle is well-known, it is because of its rich, intertwined history with the University of Toronto. The U of T administration used to refer to it as University Creek when it still flowed, and campus publications like the *Varsity* and the *Graduate* have covered stories at momentous times in the history of the creek. The deep ravine, still evident near the Faculty of Music, and now known as Philosopher's Walk, gives a sense of the intensity of the Taddle's flow. At its deepest point, just before reaching Hoskin Avenue, the valley bottoms out into a sewer grate. If you put your ear to it, you can hear the Taddle flow five feet down.

Leaving Philosopher's Walk and crossing Hoskin, Taddle Creek flowed directly under what's now the Wycliffe College tennis court and the Great Hall at Hart House, its banks becoming part of the foundation for Queen's Park Crescent West and its overpass. Historic university maps and plans show the creek was joined in places by tributaries that drained the surrounding campus lands. Landscape plans from early in the university's history were inspired by the picturesque presence of Taddle Creek, and included the creation of a pond near where Hart House sits today and botanical gardens that incorporated part of Queen's Park. (The university's 1999 Campus Open Space Plan also called for a pond to be built in front of Hart House.)

Although the botanical gardens were never realized, Taddle Creek was dammed in the early 1860s near present-day King's College Road to create the pond. Named after the first president of the university, John McCaul, McCaul's Pond flooded ravine lands further upstream toward Hoskin Avenue, creating a long, sinuous water feature on campus.

Although students used McCaul's Pond early on for fishing, skating and contemplation, it eventually came to be regarded as a public nuisance as it became increasingly polluted with raw sewage dumped into Yorkville drains. After its construction in 1881, the Toronto Baptist College (today's Royal Conservatory of Music) discharged even more raw sewage into an already-overwhelmed Taddle Creek at Philosopher's Walk, which was taken downstream by the Taddle to fester in the pond.

Despite references to the 'evil odour' associated with the pond, students wrote poems about the Taddle that indicated how important it was in their lives. The following poem was written at the time of its burial:

*University College, c. 1890.*

*O, gentle Taddle! wandering by thy side,*
*I watch thy merry waters glide,*
*And hear the murmur of thy limpid tide,*
*Taddle.*

*Of undergraduates full many a race,*
*Here by thy banks have dwelt a little space,*
*And known and loved this mem'ry-haunted place,*
*Taddle.*

*And often have thy banks and bosky glades,*
*Resounded to the laughs of youths and maids,*
*As careless, happy, free, they sported 'neath thy shades,*
*Taddle.*

*Here many a deed of blood and derring-do,*
*Has bearded Senior or relentless Soph put through,*
*And stained with Freshmen green thy waters blue,*
*Taddle.*

*But sentimental fancies, deeds of gore,*
*Shall twine around thy sacred name no more,*
*Thy days are ended, and thy glories o'er,*
*Taddle.*

*The City Council would thy stream immure,*
*And shut thee up with bricks and lime secure,*
*And make thee – Ichabod! – a common sewer!*
*Taddle.*

*Let's soothe thy parting spirit with a Freshman's blood,*
*And while there's time embed him deep in mud,*

*And sail him tenderly adown thy flood,*
*Taddle, O, Taddle!*[2]

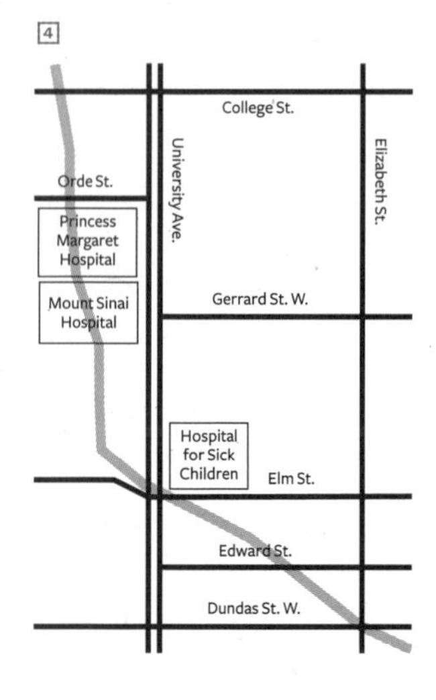


In spite of the undergraduate literature it inspired, debates about the creek's burial had begun by 1881. Much discussion ensued between the university and the City on what to do with both the pond and the creek. These discussions are reflected in various editions of the *Varsity* between 1881 and 1884, and provide much insight into how severely deteriorated the creek had become. By 1884, the pond was drained and the creek was sewered – effectively burying the last major part of Taddle.

South of College Street [4], the Taddle flowed through the lands that house the former Toronto Board of Education head offices and Orde Street Public School. These lands had originally been part of the estate of Judge William Dummer Powell, Chief Justice of Upper Canada. His estate originally extended from Bloor Street West to Queen Street West and from McCaul Street to University Avenue, lands he sold to Bishop John Strachan in 1828 for the eventual establishment of the University of Toronto. From his 'country' home at Caer Howell, located where Mount Sinai Hospital now stands, Powell and his family could see Taddle Creek meandering by.

Continuing in a southeasterly direction, the Taddle flowed through the lands of the Princess Margaret Hospital and Mount Sinai Hospital, and crossed University Avenue, cutting under where the Hospital for Sick Children now stands. It then headed through U of T's Faculty of Dentistry, and east along Edward Street to the intersection of Elizabeth and Dundas streets. Running south on Elizabeth Street [5], the Taddle turned east just before New City Hall, then ran through Larry Sefton Park at Bay and Hagerman streets to cross into an area presently occupied by the Church of the Holy Trinity and the Eaton Centre.

In the 1840s, the area around the church was known as Macaulaytown, named after owner James Macaulay, surgeon to the British Forces in Upper

2 Anonymous, *The Varsity*, October 1883.

Canada. Around this time, Macaulay began to subdivide his estate to create working-class housing, a project that was continued by his sons until 1853. The land was eventually donated to the Church of the Holy Trinity, and the Taddle did not fare well through grid-imposed subdivision there, going underground as a sewer sometime in the 1850s. From Macaulaytown, the Taddle flowed in a south-easterly direction across Trinity Square Park and through the Eaton Centre [6]. (The huge landmark fountain in the mall sits near where a bridge once carried people and goods over the creek.) It then ran close to Massey Hall, St. Michael's Hospital and Metropolitan United Church.

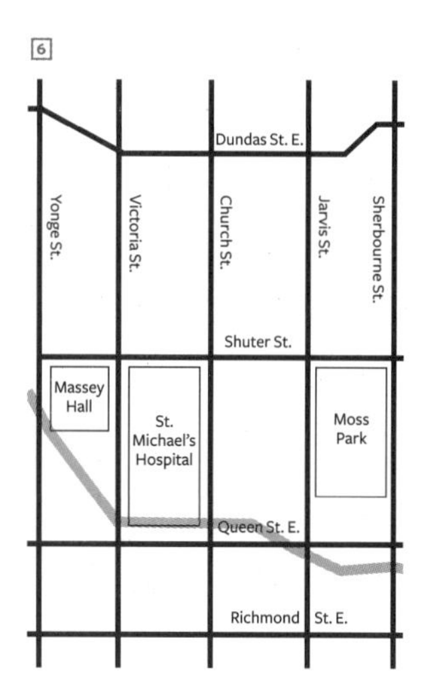


The site of Metropolitan United Church was once McGill Cottage, which sat on land that belonged to the estate of John McGill, receiver general of Upper Canada from 1813 to 1819. Amidst orchards and pine trees, the view from the cottage included the Taddle, which flowed parallel to Queen Street East. The estate itself extended 100 acres up to Bloor Street East, and included the lands that would eventually house part of Ryerson University. The estate incorporated Taddle Creek at its southern end, and also Sixth Creek, or Normal School Creek, a significant tributary of the Taddle that connected with the Taddle at Moss Park. (Moss Park still had a remnant ravine of its creek until the 1960s, when it was filled in to create the sports field that's there today.)

McGill's nephew began subdividing the estate in the 1830s, working at it well into the 1860s. Although Taddle Creek disappeared with the development of these lands, its nine-metre-deep ravine, which stretched almost from the Metropolitan United Church east to Sherbourne Street, was significant enough to pre-empt Queen Street from extending east of there up to the 1850s. At that time, in order to continue east, you had to take a detour along Britain Street, just south of the ravine, to eventually reconnect with Sherbourne Street. Indeed, the curve of Britain Street echoes the historic curve of the Taddle and its ravine. It wasn't until the subdivision

*Moss Park estate, c. 1834.*

7

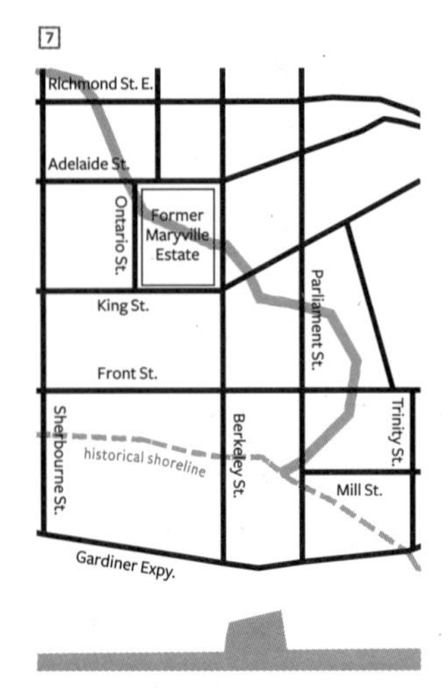


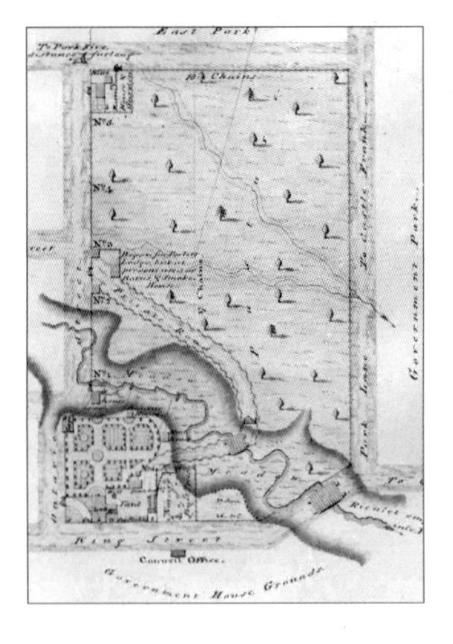

*Maryville estate, 1802.*

and development of the park lots between Yonge and Sherbourne streets that the ravine was filled in and Queen Street was extended straight across to Sherbourne Street.

As Taddle Creek cut south of Queen Street East, crossing Sherbourne Street, it entered into the Town of York's original ten-block grid 7. At Richmond Street East, near Sherbourne, it entered the town lot belonging to Thomas Ridout, surveyor general and government house member for Upper Canada. Early watercolour paintings of the property clearly show a drop in elevation – evidence of a ravine, and thus of the Taddle – in the lot. Today, a warehouse with a stepped alleyway that drops about ten feet sits on the site – a remnant reminder of the Taddle's northern ravine.

The Taddle continued southeasterly, cutting across Ontario Street, meandering along Adelaide Street East to Berkeley and then south into the Maryville estate, which belonged to yet another surveyor general and government member for Upper Canada, David W. Smith. At the southern end of the estate, at King Street East, a bridge spanned the Taddle ravine, landing on Parliament Street. From Parliament Street, the creek swung east in a large curve (the swerve of Derby Street reflects the path of the creek) to Trinity Street, then south to Front Street East/Eastern Avenue, swinging back to Parliament Street, and to its marshy mouth between Mill and Berkeley streets, where it emptied into Toronto Bay.

At the time of settlement in 1793, a marsh that gave way to a great stand of oak trees lay at the mouth of the Taddle, stretching westwardly along the shore to Garrison Creek. The deposits of sandy soil here supported the growth of these oak forests. In fact, the mouth of the creek was very important to the Aboriginal heritage of Toronto: both the Mississauga and the Iroquois First Nations still used the Taddle for social gatherings and fishing at the time of European settlement, and the Mississaugas had an encampment amidst a grove of oak trees at the bend of the Taddle, described above, where they fished for trout and salmon.

The coming of the railways in the 1850s and 1860s served to not only sever the city from its waterfront, but also hastened the disappearance of the city's creeks. The lakeshore was extended through lake-filling, piers were created and the marshy mouths of creeks like the Taddle disappeared in the process. The young city then industrialized significantly, destroying remnant natural features like valleys, hills and streams, as well as fragments of the old forest of pine and oak. It was not just Taddle Creek that disappeared dramatically under the laying of the train tracks and the establishment of the Victorian industrial age, but also its natural environment, which was fragmented into echoes of its former self – dips and turns in streets and postage-stamp-sized parkettes.

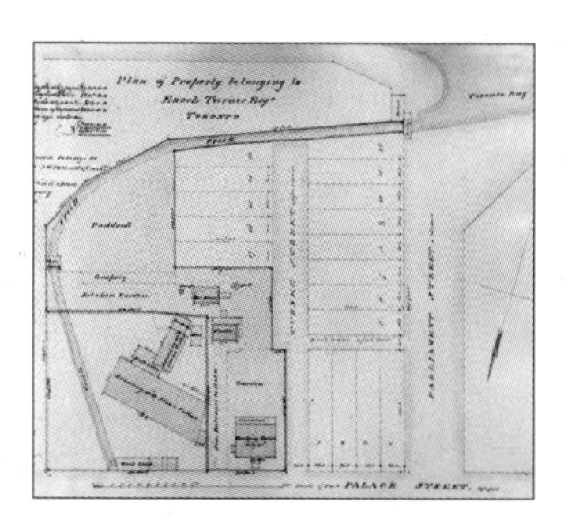


*Enoch Turner estate, 1854.*

A TADDLE TALE:
'... a stream, crystal clear ...'

'... my father decided to get a little home "in the country" and settled on a most lovely site for a cottage in "College Avenue." The ravine, which is still partly to be seen near the University, ran down through the site, behind Caer Howell ... It was crossed by a white wooden bridge, between what is now McCaul Street and the gate of my father's place. Beneath this bridge flowed a stream, crystal clear, rising in blue land on Well's Hill, and when it reached our fence ran into a large pond made with deep, shelving banks, forming a tiny island, which was a great playground for us children. Brickwork was arranged as a waterfall, over which the creek, as we called it, ran in foaming glee, and through the deep ravine, with three lovely little rustic bridges, covered with wild grapevine. My father had a regular bed made in the creek, of cobblestones and bright-coloured pebbles.'

Mrs. W. Forsyth Grant, daughter of the Hon. J. B. Robinson, January 1914, quoted in an untitled article in the *Canadian Magazine*

## LEARNING FROM THE PAST

The Taddle's transformation into a sewershed created or contributed to a host of urban environmental concerns, of which three are especially notable: poor water quality; habitat depletion and loss of biodiversity; and buried local natural and socio-cultural history.

Taddle Creek was buried primarily as a system of combined sewers, and was not designed to handle the large volumes of water the city gets after a major rainfall. As a result, an untreated, sewage-laden stew flows directly into Toronto Bay and Lake Ontario or into rivers like the Don following heavy storms, badly affecting the water quality of Lake Ontario with, for example, high counts of fecal coliform and other contaminants.

Taddle Creek was buried while its watershed urbanized. As it headed underground, all the natural habitats and features – forests, wetlands, marshes, valleys, tributaries – fed by the Taddle's watershed disappeared or were fragmented, which resulted in a serious loss of indigenous biodiversity crucial to the ecological integrity of the region.

*Girls at Wychwood Park, 1916.*

The tale of the Taddle also teaches us about buried local natural and socio-cultural history, knowledge of which is critical for re-inhabitation. It is important to understand not only the natural processes that led to the creation of one's watershed, but also the social and cultural forces that had an impact on the changes that took place. Burying creeks, eliminating habitats and impairing normal watershed functions all diminished the quality of life of the city's residents. The loss of this history impacts current and future generations, especially if they are new to this land, and can lead to living with an 'absence of place,' where, as landscape architect Michael Hough suggests, everything that can give meaning to who and what we are in the world, and how nature works, is hidden (streams, wildlife) or gone (original flora and fauna, original topography).[1] Re-inhabitating a watershed means not only uncovering the history of how the watershed was formed and transformed, but also advocating for change that prevents such things from happening again.

## ENRICHING THE LIFE OF A WATERSHED AS PLACE

> A region holds the power to sustain and join disparate people: old ground charged with common wholeness and forces of long-growing life. All people are within regions as a condition of existence, and regions condition all people within them.
>
> Peter Berg, *Reinhabiting a Separate Country*

1 Michael Hough, *Cities and Natural Process* (London: Routledge, 1995).

*Boys at Wychwood Park, 1916.*

I have written this essay to tell the story of Taddle Creek, but also to point to what Torontonians might do, and what they've already done, to re-inhabit their watershed.

On a personal level, I have tried to re-inhabit the Taddle watershed by learning its waters, finding out about its natural and social history, the role it played in the life of both Aboriginal peoples and Europeans and what happened to it over time. I have also worked with community groups along the Taddle in projects that foster a sense of pride in the community, while also providing a greater context for the creek, both at the watershed level and at the bioregional level. As Michael Hough said so well, 'to understand a local place . . . requires an understanding of its larger context – the watershed and bioregion in which it lies.' This is why I used to begin Taddle Creek walks with a discussion of the Oak Ridges Moraine and the bioregion it informs before moving into the story of the creek itself.

Meanwhile, folks up in Wychwood Park restored their pond in the hopes that it would function more ecologically. By helping to clean the water in their pond, they ultimately contribute a little to cleaning up the waters of Lake Ontario, and to regenerating habitat.

Further downstream in the Annex, people have been planting native tree and plant species that are thought to have existed when Taddle Creek flowed. Indeed, it is the Taddle's stories that keep it alive in the Annex; the identity of the neighbourhood is intertwined with the history of the watershed. Amanada McConnell, a long-time Annex resident, notes that the Taddle has frequently washed out her garden over the years, especially during spring runoff. She tried to find the creek by digging down two to three metres, but instead found an old pipe, with water still running through it, and a culvert that was most likely placed there by farmhands to drain the land of Taddle Creek. When she called City engineers to try to trace the course of the water that flowed through the pipe, they dropped a traceable dye into the pipe, but in no time lost track of it. They do not know where the Taddle waters run – at least on private property. Both she and others have also tried to create habitat for tadpoles by recreating a wetland-like feature across a few backyards, again fully aware of the presence of Taddle throughout the area.

When you head down to Trinity Square Park, you are once again on Taddle Creek. It was thus with great serendipity that the park was selected by members of the Toronto Labyrinth Community Network and City Hall as the future site of the Toronto Labyrinth. The placement was serendipitous because neither party knew that labyrinths have historically been located near bodies of water. However, once the story of the Taddle emerged, it was woven into the project so that when volunteers guide people on labyrinth walks, they weave in the story of the Taddle and the historic relationship of labyrinths to water.

Similarly, the community garden and former River of Tallgrasses at Moss Park were grown because citizens from across the socio-economic spectrum came together to create these projects at the park, not only as a form of community development, but also because they were inspired by the knowledge that Taddle and Moss Park creeks once converged in the park. At that time, a market garden had been in operation, and today, food is once again being grown there. The River of Tallgrasses was created by bringing together people from the various communities located along Taddle Creek as a way of remembering where the creek used to flow. Sadly, the River of Tallgrasses disappeared within a year due to a host of social problems, notably after needles and drug paraphernalia were found amid the grasses. It's now completely gone except for a little 'tributary' near the Moss Park Armoury made up of a bed of grasses and other native plants, including a silver maple and an American elm – species that existed when the Taddle flowed freely. All this work fostered connections between community groups along the historical course of the creek, between community groups and the Taddle itself and between individuals in the community. The projects and the programs exemplify what Stephanie Mills and other thinkers of bioregionalism and watershed consciousness are trying to get at, that re-inhabitation can start, very simply, by saying, 'I'm here, and I am here with you all.'

Re-inhabitation compels us to believe that we do not have to accept things as they are. Yes, Taddle Creek was once a living stream. Yes, its watershed has been transformed into a sewershed. But what might it be tomorrow? The stories continue to be written, or unburied, and need to be told. Re-inhabitation has the potential to ensure that those outcomes will be positive for all communities of beings that inhabit their watersheds. This idea is perhaps best expressed by the American poet and bioregionalist Gary Snyder, who said, 'Watershed consciousness and bioregionalism is not just environmentalism, not just a means toward a resolution of social and economic problems, but a move toward resolving both nature and society with the practice of a profound citizenship in both the natural and the social worlds. If the ground can be our common ground, we can begin to talk to each other (human and non-human) once again.'[2]

**2** Gary Snyder, *A Place in Space: Ethics, Aesthetics, and Watersheds* (Counterpoint: Washington, D.C., 1995).

For their knowledge and inspiration, I'd like to thank Stephen Otto, Rollo Myers, Amanda McConnell, Michael Hough, Stephanie Mills, Helen Mills, Susan Richardson and Murray Boyce.

Joanna Kidd

# How the Toronto Bay Initiative reimagined Toronto Harbour

The Toronto Bay Initiative was a citizens' group with a singular mission: to make Toronto Bay cleaner, greener and healthier. As a bare-bones, non-profit organization, TBI conducted its work thanks to the commitment of hundreds of volunteers and the support of dozens of organizations, agencies and business partners such as the TRCA, the City's parks department, Harbourfront Canoe and Kayak Centre, and Mountain Equipment Co-op. At its biggest, the organization never had more than a half-time executive director and a half-time administrative assistant. Yet, in ten years of operation, it planted over 28,000 trees, shrubs and wildflowers, naturalized 500 metres of shoreline, helped to build and maintain a new wetland on the shores of Toronto Bay and partnered in the regeneration of Toronto's only sand dunes, located on Hanlan's Point. The group also mobilized over 5,000 people in hands-on projects to improve the environment in the Toronto Bay area and educated over 3,000 people about the riches of the bay through its events.

Despite its successes, TBI no longer exists. Like many small non-profits, the organization suffered from an inability to secure sustainable core funding. The group had little trouble finding money for projects, but simply couldn't get enough funding – whether from grants, memberships or donations – to keep staff on board, pay rent on a modest office and pay the phone bill. The constant struggle to make ends meet adversely affected board members (of which I was one), stressed staff, sapped morale and distracted us from our core mission. According to the Trillium Foundation, the Province of Ontario's grant-making foundation whose raison d'être is to fund community-based initiatives, we were an example of an unfortunate trend of 'too many small, unsustainable citizens groups.' Though we had received two funding grants from the foundation, we were turned down for a third. After ten years of operation and much soul-searching, the board pulled the plug and TBI ceased to exist as of July 1, 2007.

The idea for the Toronto Bay Initiative was hatched in a meeting in late 1996 hosted by the Waterfront Regeneration Trust. The meeting, attended by then-mayor Barbara Hall, councillor Dan

Leckie and a diverse group of citizens, waterfront residents, naturalists and agency representatives, was called to explore the idea of doing something – no one quite knew what – about Toronto's Inner Harbour. People around the room spoke about their relationship to the harbour, how they viewed it, how they used it and what it meant to them. People talked about the poor water quality and the lack of public access to the water itself. People talked about the vital importance of the harbour to the founding and history of Toronto, and the fact that the city had largely turned its back on it. I remember telling the group about picking up litter along the shorelines of Toronto Island, my solo attempt to do something about the most visible symptom of the harbour's poor water quality.

Two things became apparent during that meeting. One was that people in the room believed the harbour had significant natural, cultural and economic value. The second was that no agency or organization was really looking out for the harbour in a comprehensive way. Elsewhere, agencies worked with citizens' groups to improve the health of Black Creek and the Don, Humber and Rouge rivers, but no agency or organization advocated for Toronto's harbour. By the end of the meeting, the participants had decided to create an organization focused on the Inner Harbour. And part of that decision – this was the beginning of the re-imagination – was the realization that the organization had to tackle not Toronto's utilitarian 'harbour,' but its 'bay.' And so the idea of the Toronto Bay Initiative was born, and a few months later, so was the organization itself.

Much has been written about the environmental degradation of Toronto Bay that took place as the sleepy, backwater Town of York evolved into Canada's largest city. As children in the mid-1960s learning how to sail in the bay, we used to joke about the little red worms on the bottom of the harbour, the sea lampreys that would occasionally fasten onto our sailboats, and the 'killer carp' that were, as far as we knew, the only living things in the murky waters of the Island lagoons. Unbeknownst to us, the 1960s were a watershed moment for the bay: conditions were starting to improve because of the rise of the environmental movement, the establishment of environmental protection agencies and the passage of new regulations. Bans on the worst persistent organic contaminants, improved sewage treatment, the adoption of pollution-prevention strategies and better control of municipal and industrial discharges led to improvements in the air, water, sediments and soils in Toronto Bay (and elsewhere), and subsequent improvements in fish and wildlife populations. In the early 1970s, Toronto's waterfront started its slow transformation from a derelict industrial and shipping landscape to a place for culture, tourism and residential living.

When TBI was formed in 1997, the environmental condition of the bay was a good-news/bad-news story. The bad news was that water quality was still poor, fish and wildlife habitats were limited and fragmented and, according to the parks department, tree canopy coverage in the waterfront area was less than 4 percent. The good news was that water quality had improved to the stage where fish populations were on the rebound – some thirty-five fish species could be found in the Toronto Island lagoons, for example. A number of habitat-improvement projects were being planned or had been completed in the area, and a Parkland Naturalization Program had started, in which staff worked with citizens' groups to plant native trees, shrubs and wildflowers in parks. Also good news for TBI was the fact that an established agency – the Waterfront Regeneration Trust – would provide financial and administrative support to the organization as it established itself. (For the first few years, TBI was, in fact, a project of the trust.) And best of all, there was a relatively new but rapidly growing neighbourhood along the central waterfront, a neighbourhood that was a potential source of ideas, support and volunteers. The elements were all in place for TBI to begin its work.

One of the group's earliest projects was *A Living Place: Opportunities for Habitat Regeneration in Toronto Bay*. This report, released in 1998, detailed how the bay had changed in over 200 years of settlement and provided a snapshot of its current environmental conditions. It then laid out a blueprint for how and where fish and wildlife habitats could be improved through reforestation, shoreline regeneration and the creation of underwater reefs, wetlands and wildlife corridors. Twenty-one scientists, policy-makers, engaged citizens and planners worked together on this joint vision for a healthy bay. The title was chosen to communicate two ideas: the first was that Toronto Bay should be a place where humans could coexist with fish and wildlife (the report argued that 'healthy natural systems and the functions they provide are the building blocks

*Tending the Spadina Quay Wetland, 1999.*

of sustainability'), while the second idea, embedded in the title, *A Living Place*, was a reminder that the bay was alive – a dynamic system composed of air, water, soils and living things.

Especially in its early years, TBI drew attention to the bay through its events. Round tables informed people about and engaged them in discussions on heritage, water quality, the 2008 Olympic bid, green development and other issues. Annual forums briefed people on progress made to improve the health of the bay. The most attention-grabbing event, the Big Summer Splash, drew people to the water's edge to watch citizens and triathletes as they swam across the bay from Centre Island. The group had to get special permission from the Toronto Harbour Commission to allow this to take place, as for decades no swimming had been allowed in the bay because of potential conflicts with shipping. The Big Summer Splash was held annually for four years, and included theatre, music, food and 'plungers,' brave citizens and some local councillors who jumped into the water at the Portland Slip to show their support for TBI's aims.

Meanwhile, TBI also coordinated volunteers for hands-on projects to improve the health of Toronto Bay. This included shoreline cleanups, stewardship of the Spadina Quay Wetland

and naturalization events during which TBI volunteers planted specimens of sandbar willow, red osier dogwood and eastern cottonwood beside roads, on lagoon edges, along fencelines, in meadows and on sand dunes.

Figuring that the best way to learn about the environment was to get people out in it, TBI also hosted about fifteen guided tours of the bay and surrounding areas annually. These were done on foot, on bicycle, in canoes (both big and small) and by boat. Expert interpreters helped people explore and understand the lagoons and woodlands of the Toronto Islands; the trails, woodlands and marshes of the Lower Don; the industrial heritage of the Port Lands; the nature hidden on the central waterfront; the birds, butterflies and vegetation communities of the Leslie Street Spit; and the beauty of the Todmorden Mills Wildflower Preserve. During these tours, participants were able to experience unique landscapes and see the amazing non-human inhabitants of Toronto Bay, including beavers, muskrats, foxes, black-crowned night herons, common terns, northern pike, snapping turtles and melanistic garter snakes. In every tour, the message was the same: there are some unique and wonderful places in the Toronto Bay area – please help us make them better.

Perhaps less public was TBI's advocacy work on behalf of the bay. Over the years, board members and volunteers sat on dozens of committees and participated in many planning processes. In so doing, they helped shape the City's Sewer Use bylaw, its new Official Plan, the Wet Weather Flow Management Master Plan, the Bike Plan, park and street design in the Harbourfront area and many of Waterfront Toronto's planning processes, including the project to create a new mouth for the Don River.

TBI leveraged its activities by tapping into the expertise and energy of individuals and partner organizations. Through partnerships with the University of Toronto and Ryerson University, TBI was able to harness the creativity of planning and environmental studies students on projects relating to green development on the waterfront, the creation of a fish habitat at Maple Leaf Quay and the mapping of the riches and uniqueness of Toronto Bay.

It is still sad to think of the loss of TBI. But eleven years after its formation and a year after its demise, there is cause for cautious optimism about Toronto Bay. Over the next twenty-five years, the Wet Weather Flow Management Master Plan will dramatically reduce the discharge of stormwater and combined sewage to the bay and yield much-improved water quality. The agency coordinating waterfront regeneration, Waterfront Toronto, is developing an enhanced network of parks and open space along the water's edge. This will increase public access to the area and tree canopy coverage, improve connections between natural areas and enhance wildlife habitat. The creation of fish habitat along the north shore of the bay will support burgeoning fish populations. The 10,000 people now living along the central waterfront, and the 40,000 or more the City predicts will move in over the next twenty or so years, are potential users and stewards of the bay, neighbours who can be tapped to appreciate, care and look after it.

*Across Toronto Bay by ferry to Toronto Island, 2007.*

Some of TBI's programs continue: Evergreen has taken on its planting program, and Citizens' Environment Watch its water quality and beach work. What has been lost, however, is TBI's education, interpretation and advocacy functions. Toronto now no longer has an organization that thinks and dreams about the bay, one that helps people explore and understand it, one that advocates for a cleaner, greener and healthier Toronto Bay.

There is still much to be done to return Toronto Bay to health. There are trees to plant and pollution to control. There are wetlands to build and fish habitat to construct. There are shorelines to naturalize and stabilize. There are municipal and waterfront revitalization policies, programs and plans to shape and politicians to nudge. This kind of stuff doesn't happen on its own. Making sure it happens will require that all levels of government make the bay a priority for monitoring and remediation. It will likely also require the genesis of a new citizens' group devoted to the bay. And what would that organization look like? If we had to do it all over again, I'm not sure we would do it very differently, except to be more strategic in our fundraising. TBI has left behind a legacy – a vision of a healthy Toronto Bay and a road map of how to get there. It is up to the rest of us to ensure that this promise is fulfilled.

Mark Fram

# A tale of two waterfronts

Toronto and Chicago: two metropolitan sprawls with several million inhabitants apiece, each addressing its respective Great Lake as if a great sea, its far shore lost over the horizon.[1]

It's instructive to compare how each city has treated its waterfront over the past century and a half, and many commentators have done so. Notably, during the 2003 debate over a proposed bridge to Toronto's Island Airport, much was made in the Toronto press of how Chicago mayor Richard Daley attended to his own waterfront airport 'problem' by sending in bulldozers at midnight to dig up its runways. In retrospect, it was a barely concealed playground taunt aimed at Mayor David Miller's newly minted administration, challenging it to show similar 'nerve.' Years later, there remains a certain muted tension in Toronto's chattering classes about Mayor Miller not quite living up to the sweeping-broom theme of his first mayoral campaign.

So. Given the title of this piece and that set-up, you'd be looking for another city-versus-city slugfest. But you'd be wrong.

The two-waterfronts title has an apt, if worn, literary resonance, appropriating Dickens like some shaggy-dog story journalists trot out on a slow news day. A hackneyed, attention-grabbing conceit. But this is indeed the tale of two waterfronts – it's just that both are in Toronto. In the very same place. Toronto would do far better by looking less dreamily or enviously at Chicago and a little more carefully – and ingeniously – at its own geography and history. This is about Toronto versus itself.

The setting for the twin tales is the 'central' waterfront, between the mouth of the Don River and the channel at the harbour's western end. Specifically, the spotlight is on the eastern parts that are, at this writing, about to be, one way or another, 'revitalized.'[2]

1 Minor correction: it's now the twenty-first century and, with binoculars, you *can* see the opposite side of the lake from a high-level resto-bar.

2 Harbourfront, to the west, has already been done, in more ways than one.

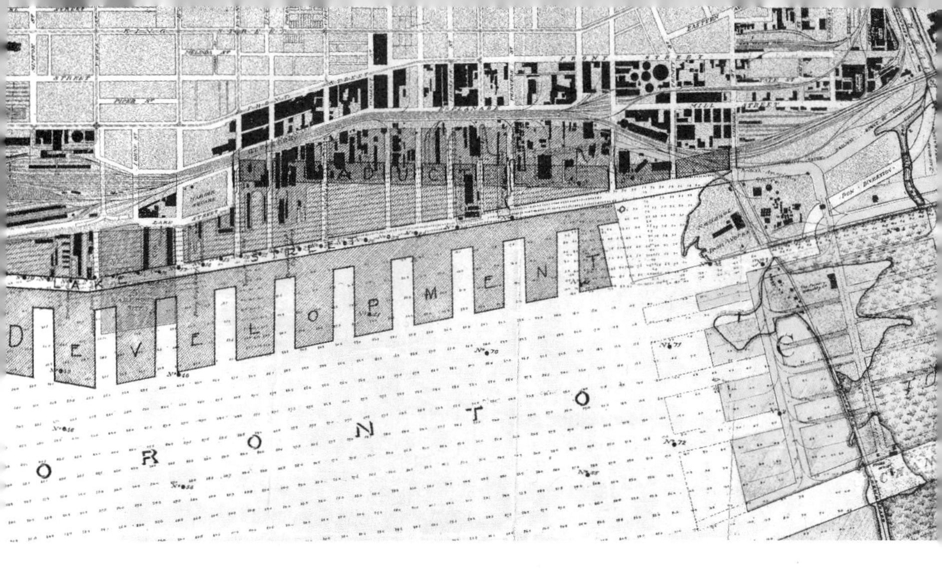


## THE COUNTERFACTUAL SHORELINE

East of Yonge Street, there is the waterfront you can see and, beneath it, evidence of what was, and what might yet be.[3] Regrettably, the truly transformative potential of the water's edge is nothing like the recently announced visions for the waterfront.[4] If executed as officially promised, the collection of new office blocks, yet more residential condominiums, a tightly packed college campus and pocket parks only incidentally related to the water will confirm that the woebegone norm of the twentieth century will be the template for the twenty-first.[5]

These impending civic actions (well, plans for action) to 'enliven' the eastern part of the harbour have been a very long time coming – perhaps because of the extended recovery period needed after earlier civic action, begun a century ago, that buried the docks, snatched much more land than needed from the murky waters and then hid it all behind an embankment, thereby enabling decades of lifelessness.[6]

Though Toronto's location was determined by its sheltered though shallow harbour, and though the city was, among other things, a modest working port well into the twentieth century, the real water's edge was incidental to the city's original status as a British colonial capital. Even now, the harbour

*The Toronto Harbour Commissioners' 1912 plan of works, an elegant rendering overlaying the ambitious lakefill onto the old waterfront and signed by, among others, the eminent landscape planner Frederick Law Olmsted Jr., remains one of Toronto's most unusual documents: a grand plan actually carried through as intended. But never really* finished.

3 'Counterfactual history, also sometimes referred to as virtual history, is a recent form of historiography which attempts to answer "what if" questions known as counterfactuals. It seeks to explore history and historical incidents by means of extrapolating a timeline in which certain key historical events did not happen or had an outcome which was different from that which did in fact occur. The purpose of this exercise is to ascertain the relative importance of the event, incident or person the counterfactual hypothesis is negating. . . .' See en.wikipedia.org/wiki/Counterfactual_history.

4 www.waterfronttoronto.ca/

5 See 'A note on terminology,' overleaf.

6 What is there to say about the Guvernment? Lively, in its way, but behind closed doors and only after dark.

remains incidental to Canada's biggest city, despite numerous well-laid plans and almost shameless civic puffery.

So, in place of this century-old continuum of hopeful plans and stillborn results, please consider, briefly, Toronto's counterfactual waterfront: an alternate world beneath the pavement. Beneath today's flatland lie the ingredients for a more genuine and provocative civic waterfront.

This alternative world is made up of a series of shorelines, the contours of scores of little lakefills. These older shores begin at Front Street, now a noisy, dirty, ten-minute walk from the water (or from a chain-link fence overlooking the water). Geomorphological and archaeological burials testify to the fundamental tragedy that has been long hidden from full scrutiny: Toronto has never regarded the shoreline of its bay as a waterfront. The place where land meets water was and remains entirely a land-front. The city has pushed the water's edge as far away as possible.

A litany of stories about the barriers between central Toronto and the harbour fill innumerable pages of civic histories, press accounts, governmental reports and academic studies: the mid-1850s sell-off to the Grand Trunk Railway of the public promenade, promised in 1818 but not delivered; syncopated cycles of the small-scale expansion of docks and warehousing into the bay; the wholesale lakefill meant to create a full-blown industrial district; the elevation of the railways on a massive earthwork; the widening of two parallel arterial roads; and – what still seems to some the ultimate indignity – the construction, in the late 1950s, of an elevated highway, the Gardiner Expressway. (But more about that later.)

Throughout much of the nineteenth century, what now lies between the earlier Front Street and today's front at the water was a mix of sand beaches, muddy bluffs, ramshackle wooden and stone piers, schooners, rafts, marine liveries, boatyards and all the other bits of a modest working harbour. By the

A NOTE ON TERMINOLOGY

Waterfront Toronto's previous incarnation was the more prosaic Toronto Waterfront Revitalization Corporation. 'Revitalization,' eh?

It is no longer possible to refer simply to, say, urban development, or even 'building.' The makeover of urban vocabulary at the keyboards of journalists, corporations and academics is certainly not confined to Toronto, though this trend of trends is most evident in metropolises and megalopolises that are growing and generating wealth, rather than in those urban locales – mere towns and cities – currently grasping at almost any new investment (or, more pointedly, reinvestment). In so-called megacities like our own, even the smallest construction acquires its own special thematic (that is, political) tilt, whether promotional or pissy: neighbourhood stabilization, social integration, cultural harmonization, mall-ification, sprawl, intensification, sustainability or, that most agonized tag of all, gentrification, a term coined to describe the ways that declining residential areas recover an earlier state of well-being by pushing out the riff-raff. Lately, entire non-residential, down-on-their-luck districts have been demolished and replaced with faux-Victorian row housing; that too has been tagged as gentrification.

Early-twentieth-century urban texts invented, even celebrated, a pejorative stew of urban ills, slums, decay and, most of all, blight (the wallow faintly recalled by a different political faction later in the century under the headline *revanchism*). But such 'critical' attempts to define urban situations have not been able to hold their own against the promotional dictionary of real estate, which has grown from a modest commercial service into a behemoth industry (the term *realtor* is not in fact a mere word, but a registered American trademark). My recent favourite is *cottageinium*. (Go ahead, Google away!)

The critical geographer trumpets gentrification as a battle cry against the forces blowing apart older neighbourhoods, while the sales agent replies, 'I *love* gentrification!' and pockets her commissions. And so, revitalization – even mere *improvement* – is hardly ever what it seems, let alone innocent.

*Today's lands, 1852's waters.*

end of the nineteenth century, the working harbour had gone through periodic reconstructions and extensions that produced dozens of piers, multiple fingers of docks, workshops and warehouses – the whole thing embraced in a tangle of hulls and masts and motion. It was busy, messy and not very pretty: no working waterfront is a safe tourist attraction up close, and decaying working waterfronts are very often civic nightmares in many ways. Any number of films noir from the 1930s on – with lurid tales of smuggling and dumping by night, gangsters on the run, dockland violence and civic corruption, pock-marked by tragic deaths – paralleled the cycles of official reports clamouring for cleanups.[7]

The nineteenth-century working waterfront was buried not long after the beginning of the twentieth, the actual work beginning during the economic

7 Ah, now the case for the Guvernment becomes clear: what could be a more genuine heritage feature for the waterfront than a speak-easy holding 2,000 or 3,000 people? A casino, perhaps . . .

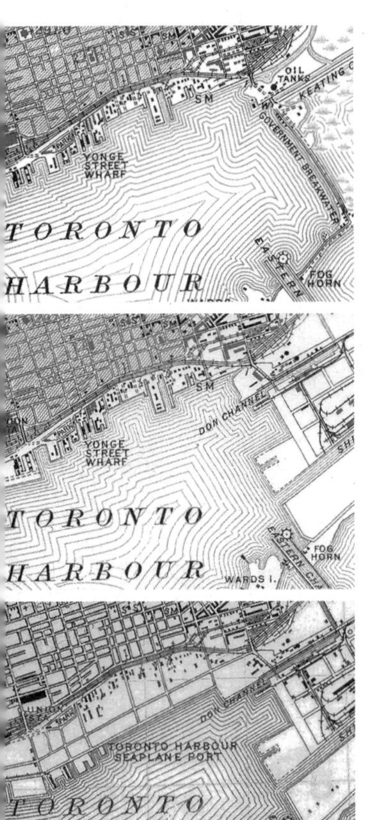


*The harbour as it was: 1909, 1918, 1931.*

boom that occurred just before the World War I, but stretching out for almost a half-century more.

What officials now call the East Bayfront – where you see the last vestiges of that ambitious harbour planning begun in 1893, climaxing after the creation of the Toronto Harbour Commissioners in 1911 and moribund ever since – wiped out those earlier messy waterfronts. The piers and slips were filled and covered with the dirt and rubble excavated from tens of thousands of basements and sewer tunnels, as well as with chunks of demolished buildings and immense volumes of sand sucked up from the harbour itself.

The new terrain was meant to be a thoroughly modern twentieth-century port and factory district, pushing eastward what remained of the evicted working port itself and offering reclaimed land right across the tracks from the central business district. The result was, and remains, a flat and sorry lakefill, detached from the working city, and with hardly any working or even accessible frontage on the water unless you're a very hardy sailor. The only real port industry that remains is the Redpath sugar refinery of 1957.[8]

To be charitable, you might allow that the harbour project took much too long to start, suffered interminable delays through wartime and economic depression and was simply never properly completed. The unfulfilled result is only better known than what's buried because it's all we see. It's all the city has seen for almost a century. All land, not much water.

The failure to allow for more than the merest carefully rationed trace of interchange[9] where land and water meet in east-central-waterfront redevelopment – whether through lack of foresight, lack of nerve, technical inability or downright refusal – condemns the East Bayfront to flatness and sorriness for many more decades, if not forever. Distant views of water from the upper storeys is all anyone can hope to have as a result of the currently planned condos and offices, no matter how inventive (or not)

8 Indeed, despite its relative youth, Redpath Sugar is actually a *heritage* site. www.hpd.mcl.gov.on.ca/scripts/hpdsearch/english/popupSearch.asp?pid=11834

9 That is, where the water shapes the land no less than the land shapes the water. As evident throughout this book (at least), this can never be a simply 'natural' process.

their architects. Certainly no dock space or boat space, never mind swimming space. Compared to even the smallest docks and inlets of the 1800s, the few tiny protrusions evident in some of Waterfront Toronto's recent renderings seem extremely little, and extremely late.

*Today's lands, 1899's waters.*

It is at this point that Toronto needs to shed its blinkers and look more clearly at other 'revitalized' waterfronts popping up worldwide. If the city can let go of the idea of making a boxy sarcophagus to cover the existing burials, it might discover that what it once buried is what other cities are celebrating.

*The harbour of the 1890s, now entombed well beyond the furthest vessel in view.*

The alternative? Bring the water back into the flatlands. Recover some of those older, multifarious dockland edges. The possibilities of both design and economics are almost absurdly simple to calculate, even if you consider them in purely numerical terms. The longer the edge of the water, and the more inlets, dock space, entry and exit points, etc., the more opportunity there is for interaction between land and water. Which, I'm sure most people would agree, is normally the whole point of a public waterfront.

Attractive active public waterfronts are rare. The shorter the linear interface you get when you arrive, the more rationed the water's edge is, with the fewest kinks or corners and a resolutely minimal boundary between land and water drawn with a straightedge – the fewer public places and activities it can accommodate. Why is Toronto's so rationed, so relentlessly 'modern'?

Cheaper to make and more defensible, presumably. Fewer cafés, with much higher prices? Perhaps it's that the minimalist straight-line dock edge is less hazardous, representing the triumph of insurance over human interest. Or was the 1912 water's edge kept as a simple surrender to the engineering technicalities of standing and moving water? Ah, so Toronto. Don't we deserve better? Is the vision for the harbour circa 1912 really so worthy of that kind of preservation?[10] The minimal edge is more controllable, and thereby more controlled. Even if it is publicly accessible, a narrow boardwalk is a mere token of the possibilities. It might belong to everyone in rhetoric, but offers so little room to move in reality.

Recently, isolated archaeological digs on now-landlocked sites have brought up plenty of harbour jetsam, some of it collected and stored for some museum that doesn't yet exist,[11] the sites themselves re-encased with nary a trace within and beneath the foundations of new buildings.

Those digs are not voluntary heritage explorations; they are actually conditions of approval for new development in this part of the city. But with a bit more foresight and effort beyond last-minute rescues, these excavations might be more: they could be tiny steps toward the recovery of the real thing, the long-buried edge of land and water some way inland from today's concrete walls.

So dig out some slips and let them fill with water. It's as simple – and unlikely – as that. Maybe not so

**10** Even Harbourfront, oft-maligned as another barrier between city and water despite its public spaces and water edges, offers some modest bits of real land-water interaction, though far short of those of its cousins at Montreal's Vieux-Port, or Vancouver's Granville Island. The small boat basin and pedestrian bridge, a tiny artistic wetland and the barely accessible inlet covered by four lanes of Queen's Quay West? Nice touches, yes (added to more capacious marina parking further west), but really a bit of a tease compared to Harbourfront's promise in the 1970s; though, to give them their due, the landscape/waterscape fragments do offer a glimpse of possibilities more substantive than the nearby and more recent faux beachfront of HTO Park.

**11** Toronto certainly does not lack for museums, galleries, archives and the like, large and small. Indeed, the small and modest locations are often the most engaging and enlightening. But since the 'provincialization' in 1966 of the former Art Gallery of Toronto, the absence from the repertoire of a big-city historicocultural destination – like Chicago's Historical Society, for one – has weighed heavily, though intermittently. The project for a Toronto museum on the waterfront is at least two decades old but still kicking, the most recent proposition located at the Canada Malting silos and docks (which had previously been touted for a music museum). Meanwhile, a selling point for a new condominium tower atop the former O'Keefe Centre on Front Street (the original waterfront) is a supposedly public museum space called the 'Arts and Heritage Awareness Centre' whose yet-unknown displays or programs might actually compete against rather than cooperate with the waterfront location. Still, with a will, and a few more millions, there ought to be a way: the City has been officially warehousing stuff since 1881, so there should be lots to show. See www.toronto.ca/culture/historical_collection.htm.

*Today's lands, 2020's waters?*

cheap, but wouldn't new water edges make the area worth a lot more than what's there now? Create hundreds, maybe even thousands of metres of new water edge by reopening some land/water fingers, by digging out the filled-in piers and making new ones on the old model. New waterfrontism!

How to pay for it? Well, most so-called realists would think this hugely expensive. But a few might notice that it could also be immensely lucrative. If, for instance, a house lot (yes, *sans* house) in the Annex is closing in on a million bucks, how much might a waterfront lot in central Toronto sell for? Do the sums, create a few hundred such sites, sell off a bunch while keeping the best of them for public access, and make an international-style investment in the kind of water's edge right here that landowners and governments of every stripe have refused, for the last century and a half, to contemplate.[12]

**12** Consider this 'challenge' from *Monocle* magazine (issue 15, volume 02, July/August 2008): 'Toronto, Richard Florida's new home, has all the assets to lead a quality of place revolution in the Americas – good international flight connections from the city's main hub and short hops from its downtown airport, a well educated, diverse population and a thriving city core. What's the problem? Toronto suffers from a severe case of the "comfies" and needs to up its game. Hopefully, Florida and his host school can place a rocket in the right place.' Sounds like an international playground taunt, doesn't it? Toronto doesn't make it to *Monocle*'s Top 25 liveable cities, but Vancouver (#8) and Montreal (#16) do. Mind you, the *Monocle* crew might not get along with official Toronto, in that it considers the Island Airport a definitive feature rather than a bug.

*Suppose the Gardiner was a structure waiting to be filled in and filled up, the harbour's upper deck . . .*

THE GARDINER PROMENADE

Now, back to the Gardiner. The idea of demolishing all or part of the Gardiner and replacing it with an eight- or ten-lane boulevard as a way to reconnect city to waterfront has been gathering steam for several years. A multi-million-dollar and multi-year environmental assessment has begun.[13]

Despite the relentless rhetoric about how the elevated roadway and ramps 'block' the city from its central waterfront, the Gardiner's removal would do little more than allow for better postcard shots of the graffiti-covered freight cars parked along the even wider Lake Shore Boulevard. The Gardiner is actually the least of the real barriers to movement between the old front (that is, Front Street) and the current front. The railway embankment of 1923 – part of the same engineering zealotry that accomplished the harbour lakefill after 1911 – permits tunnels at only four major streets east of Yonge, broadcasts train noise for blocks at all hours and denies views north or south unless you're above the third or fourth storey of any neighbouring building (apart from weed-covered gravel hillsides and concrete or steel retaining walls).

13 There is already the six-lane Lake Shore Boulevard and the four-lane Queen's Quay, so how could the stroll north/south, downtown-to-harbour, get any worse? The preliminary controversy about 'impact,' the issue driving the environmental study, is rather about the drive time from the east end to the downtown office.

*. . . overlooking, well, no longer this kind of activity perhaps (in this case, the waterfront promenade of Algiers at the end of the nineteenth century), but some real twenty-first-century life and purpose on the water's edge.*

Reality check. The Gardiner is troublesome as a roadway, but it isn't much of a north/south barrier. Whether or not motor traffic is above ground or on the ground (or, at colossal expense, underground), the route still carries a lot of vehicle traffic just passing through. Yes, a few productive lands are catered to by its route: a pack of office towers, a web of streets and underground trails, a couple of arenas and even an ex-distillery, all also extremely well-served by public transport. The Gardiner also offers a reasonably convenient way out of town for the rapidly expanding population of waterfront condo owners. Removing the structure might kick up some of their property values, in return for which other neighbours, and the city as a whole, would gain precious little. There would still be only eight narrow sidewalks in four grotty passages between the city grid and the lure of the lake.

Instead, consider that the Gardiner – from both east and west into downtown – offers one of Toronto's especially rare dramatically scenic routes: a lake and bay on one side, a dynamic mountainscape of concrete, steel and glass on the other and, at the very centre, the intricacy of the urban core. It's actually

*Where the water used to meet the land.*

one of the few exhilarating direct downtown entries anywhere in the world. If cars can't have it, why not hand it over to bicycles, joggers, walkers and all the cafés, gardens, vistas and street life it can handle? Not to mention the few hundred front doors to those inevitable condos, at a new level, above the traffic.

To demolish and discard the big ugly concrete ribbon would cast aside the structure's immense potential as a world-class promenade. Remember when people had delusions of a world-class Toronto? Well, here's one way to bring 'em back. Toronto needs something more than a mere boulevard. Think of the Gardiner as a new street level both overlooking the water and enabling pedestrians and bicycles to amble over the railways at many more crossings than exist. In that second, simple big stroke, the City could further redeem the flatness and sorriness of the Harbour Commissioners' lakefill.

#### NEVER MIND THE DON: BRING BACK THE *HARBOUR*

Toronto's orgy of envy at American pyrrhic expressway demolitions – like that of Boston's Big Dig, or the refit of San Francisco's Embarcadero – is truly misplaced. Those are very different cities, very different waterfronts and downtowns, and very different fiscal regimes. Their projects restored some

real connections – and added big public spaces – to what had remained rather more real and currently operating waterfronts than Toronto's. Both attracted massive investment (and debt) from every level of government. And apart from removing ill-considered roadways, neither project pretended to mask a previous century's worth of flatness and sorriness.

The return of real pieces of Toronto's waterfront to the city it spawned, and an elevated public space and promenade to relish the new/old views – you might call them hometown re-visions rather than foreign imports – are rather adventurous, shall we say. Toronto, as you are undoubtedly aware, might well be regarded as the epicentre of the unadventurous. But isn't the counterfactual waterfront just a teensy bit more tempting than what promises to be little different from a Harbourfront East, but with more buildings and fewer boats?

So here's a new dare to the mayor and to the people who dropped 'revitalization' from their name. Here's your chance to redeem the flatness and sorriness of the East Bayfront before it is entombed for another sad century.

Bring the waterfront back to the city. And give us an upper deck to watch the action.

*The fish market: two views, 1842.*

Jennifer Bonnell

# Bringing back the Don: Sixty years of community action

In the summer of 1983, Charles Sauriol, conservationist and long-time champion of the Don River Valley, sat down to capture some of his memories of the valley in the 1920s. He recalled a time before the Don Valley Parkway punched its way along the valley bottom, before the worst of the urban pollutants fouled the waters of the river – a time when the river valley was still largely rural, and partly wild. Thinking of the summers he and his family had spent in a cottage at the Forks of the Don, he wrote:

> I remember seeing the full moon break over the pines, spreading its beams of mysterious phosphorescence over the misty shrouds that rose from the river to the flood plain... Many an evening I walked to and from the swimming hole as twilight gradually closed down on the day. Then, seated in front of the cottage, I could hear the water flowing over the river stones, and sometimes, just at dusk, the strident call of a whippoorwill.[1]

Sauriol's love for the valley was rooted in personal experience – in time spent living and playing in the valley, tending his vegetable garden, harvesting honey from his apiary, watching his children swim in one of the river's rare deep pools. Biking through the valley's cool green corridors on a summer evening, or taking in the sweep of brilliant fall foliage from the Don Valley Parkway, many of us have felt something like Sauriol's deep sense of place, and of fortune that such a unique urban wilderness exists in the heart of Canada's largest city. The valley, with its resilience through more than 200 years of human settlement and modification, spoke to Sauriol, as it does to us, of hope. Hope for a new commitment to accommodate and nurture natural systems as our city continues to grow, and hope that the legacy of past mistakes, borne in the landscape of the river, will teach us a new way of belonging in the very specific place where we live. For Sauriol, lived experience in the Don Valley led to a lifelong quest to protect it; by drawing together and building from our experiences, we can begin to do the same.

1 Charles Sauriol, *Tales of the Don* (Toronto: Natural Heritage/Natural History, 1984), p. 19.

## CHARLES SAURIOL AND THE DON VALLEY CONSERVATION ASSOCIATION

Charles Sauriol's life story is also the story of the Don River Valley. In his youth in the 1920s, Sauriol developed a passion for the outdoors, camping in the woodlands of the Don as a member of the 45th East Toronto Boy Scouts Troop. Over the years, he refined his skills as a naturalist by cataloguing the plants and animals – muskrat and leopard frogs, white pine and basswood – he observed on his frequent rambles through the valley. He was also painfully aware of the dramatic changes that had occurred in the valley. His father, Joseph Sauriol, had played an intimate role in the reconstruction of the Lower Don in the 1880s, operating one of the dredges that carved a new path for the river and replaced its meanders with a straight, hard-edged channel. But Sauriol gained his real knowledge of the Don, and the heartbreak that came with that knowledge, during the forty-one summers he spent with his family at a cottage at the Forks of the Don, near Don Mills Road and Lawrence Avenue East.

*Charles Sauriol in front of the original cottage at the Forks, July 1935.*

Every May from 1927 until 1968, Sauriol and his family moved from their city home to a rustic cottage in the valley, foregoing electricity and indoor plumbing to live close to the land. They were summers that, as Sauriol wrote in his 1981 book, *Remembering the Don*, 'filled my time with the orchard, the garden, the apiary, the easy living by the then clean Don River.' Forced to vacate the cottage in 1958 to make way for the Don Valley Parkway, Sauriol and his family moved across the Forks to the Degrassi homestead on the East Don. Their time in the valley, however, was almost up. Widespread damage caused by Hurricane Hazel in 1954 gave new urgency to flood-control measures, and the newly formed Metropolitan Toronto and Region Conservation Authority wanted to remove houses from risky flood-plain areas. In 1968, the Sauriols were expropriated for a second time, and this time they left the valley for good, reconstructing their summer home on a property near Tweed, Ontario. In his time on the Don, Sauriol had seen the valley change from a picturesque setting of rural farms and woodlands to an increasingly threatened corridor of urban green space.

By the 1940s, the Don was no longer the same river that Elizabeth Simcoe, wife of Lieutenant-Governor John Graves Simcoe, had described in 1794 as 'abounding with wild ducks & swamp black birds with red wings.'[2] Straightened and industrialized in its lower reaches, the river had also borne

2 Mary Quayle Innis, *Mrs. Simcoe's Diary* (Toronto: Macmillan, 1965), p. 104.

the impacts of deforestation and residential development upstream. With the draining and filling of the Ashbridge's Bay marsh in the early 1900s, the river had lost a crucial absorption and filtering function. Silt from upstream development flowed downriver and accumulated in huge quantities at the mouth of the river, hindering navigation and propelling the Toronto Harbour Commission to undertake costly annual dredging to keep shipways clear. Pollution had also plagued the Lower Don since the early decades of European settlement; while raw sewage and cattle manure had been the main sources of river pollution for much of the nineteenth century, rapid industrialization in the early decades of the twentieth century added new and dangerous effluents to the cocktail, including oil by-products released by the refineries at the mouth of the Don and chemical wastes from the soap factories and paper mills further upriver. Across the city, population growth pushed outdated sewage infrastructure beyond capacity, and increasing amounts of partly treated sewage were discharged into the city's river systems. Tests by the Provincial Board of Health in 1949 found a daily average of 6,500 pounds of suspended solids dumped into the Don River from six sewage treatment plants – almost double the normal summer flow of the river itself. Conditions became so bad that, in 1950, a provincial conservation report described the Don as an 'open sewer' and ranked its water as the most heavily polluted in Ontario.[3]

Perhaps the biggest threat facing the watershed, however, was urbanization. In the years following World War II, more and more valley lands (and the adjacent tablelands that drained into them) were earmarked for residential and industrial development. In 1949, Shirriff, the jam and preserves manufacturer now owned by the American firm Smuckers, proposed to construct a factory on the site of Todmorden Mills, a proposal that would have seen the existing historic mill and brewery buildings demolished to make way for a storage facility. Community opposition to the proposal was swift and vocal, and Shirriff abandoned the project the following year.

One of the outcomes of this successful community-based campaign was the formation of the Don Valley Conservation Association, established by Fantasy Farm owner Rand Freeland, East Toronto lawyer Roy Cadwell and Charles Sauriol in November 1949. For the next eight years, Sauriol and the DVCA worked to protect valley resources and educate the public about

3 Ontario Department of Planning and Development. *Don Valley Conservation Report* (Toronto: Queen's Printer, 1950), Part VI, Chap. 3, p. 15.

DVCA *Conservation Special departing Don Station, c. 1954.*

the need for conservation. Members of the DVCA patrolled the valley to protect trees from the hatchets of young boys and rare wildflowers from the enthusiasm of their admirers. Nature walks and annual tree-planting days helped inform the public about the threatened wilderness at their doorsteps, and the first Paddle the Don event, organized by the DVCA in 1949, encouraged Toronto residents to see the valley as a place for fun and recreation. In 1951, Sauriol organized the first of eleven popular steam locomotive trips – what he called the Conservation Specials – through the valley and beyond. The trip originated at the Don Station on Queen Street East and retraced by rail Simcoe's journey to the headwaters of the Don in Richmond Hill in 1793. High school students in period costumes adopted the roles of Simcoe, his wife Elizabeth and his aide-de-camp. The events were a huge success, attracting over 800 people to support the conservation cause.

At the same time as these initiatives, farmers, naturalists and foresters across Ontario were expressing growing alarm about the effects of soil erosion and flooding. In response to these concerns, the provincial government passed the Conservation Authorities Act in 1946, which enabled local residents to request a conservation authority to manage resources in their watershed. Two years later, the Don Valley Conservation Authority was established – confusingly adopting the same acronym as Sauriol's DVCA. The authority differed from Sauriol's grassroots group in its access to funds from the Province and municipalities in the Don watershed, and in the technical support it received from the Ontario Department of Planning and Development, which published the comprehensive *Don Valley Conservation Report* as a background and guide for conservation activities in the watershed in 1950.

After joining the authority as the leader of its East York Branch in 1954, Sauriol built upon many of the grassroots activities he had initiated six years earlier as founder of his DVCA. Then, on October 15, 1954, Hurricane Hazel hit. Although no lives were lost in the Don Valley, two cars and their occupants were swept into the river; one man waited over eight hours in an elm tree before being rescued by authorities.

As the city rebuilt over the winter of 1954–55, it did so with a new awareness of the significance of valley lands as natural drainage channels for flood waters. In 1957, four Toronto-area conservation authorities, including the Don, amalgamated to form the Metropolitan Toronto and Region Conservation Authority, which allowed for greater coordination between jurisdictions in regulating the use of valley lands. The MTRCA had the power to acquire valley lands for flood control and recreation purposes, a decision that would have dramatic consequences for the future of the Don Valley. Charles Sauriol played a large role in these acquisitions as chairman of the MTRCA Conservation Areas Advisory Board from 1957 to 1971, and as the first executive director of the Conservation Foundation of Greater Toronto – the fundraising arm of the MTRCA – from 1963 to 1966. Between 1957 and 1994, approximately 15 percent of remaining natural areas in the Don Valley were saved as part of the MTRCA flood-plains protection program. At the same time as the MTRCA began acquiring valley lands in the late 1950s and 1960s, the newly formed Metro Toronto Council made massive investments into the city's aging and overburdened sewage infrastructure, closing down small, inefficient sewage-treatment plants across the city and constructing trunk sewers through the major river valleys to carry flow to new and expanded sewage-treatment plants on the lakeshore. These developments had profound implications not only for river-water quality but also for the enjoyment of newly created valley parklands once made unbearable by the stench of sewage.

While protection of valley lands from private development and sewage pollution were important milestones in the conservation history of the Don, much remained to be done. In the years after Hazel, the MTRCA's focus on flood control tended to emphasize engineering solutions such as dams and channel reinforcements over habitat protection and environmental restoration. Groups like the Toronto Field Naturalists continued to press for more comprehensive ecological protection for urban valley lands, however, and in the following decades, a new generation of environmental activists began to lament the ongoing pollution of the river by stormwater runoff and other sources.

*Don Valley safety officer Robert Speakman and Al Comber feeding three-to-four-week-old raccoons.*

### GROWING PUBLIC AWARENESS, 1969–1989

By the late 1960s, despite improved sewage treatment and significant reductions of sewage flow in the Don, the river was still dangerously polluted. Local industries continued to discharge harmful effluents into the sewer system, and combined sewers in the older parts of Toronto, including most of the Lower Don, continued to overflow during periods of heavy rain, sending raw sewage into the river. The river had also become increasingly inaccessible to Toronto residents, especially in its lower reaches. The construction of the Don Valley Parkway and the Bayview Extension in the late 1950s and early 1960s had cemented the perception of the Lower Don as an urban wasteland criss-crossed with rail and road arteries and littered with abandoned industrial buildings, road-salt storage sites and equipment storage yards. Highway construction destroyed a key wildlife corridor in the valley bottom and redirected the river's water flow, contributing to its low, listless appearance. Fences erected along the freeways made public access to the lower river valley very difficult, further sealing the fate of the Don as out of sight, out of mind. It had become very difficult indeed to imagine this portion of the river valley as settler William Lea remembered it in an address to the Canadian Institute in 1881:

> This wooded portion of the river was one of the most beautiful walks that could be taken. Here was quiet, only the rippling of the water over a stoney bed, or the whirr of wild ducks, or the partridge drumming in the distance. The water was pebbly and clear, the banks covered with evergreens and trees, forming a canopy of beautiful green. A temple not made with hands.[4]

4 Printed in the *Toronto Evening Telegram*, February 4, 1881.

*Pollution Probe's 'Funeral for the Don,' 1969.*

In 1969, Pollution Probe, an ad hoc group of University of Toronto professors and students led by zoology professor Donald Chant, brought the plight of the Don to public attention. Declaring the river 'dead' as a result of years of pollution and detrimental development, they donned clothes of mourning and led a funeral procession – Chopin's *Funeral March* playing in the background – to the banks of the river. The funeral event received widespread media coverage and fuelled new demands from individuals and community-based organizations for a cleaner and more accessible Don. Pollution Probe continued its campaign with educational tours of the Don that demonstrated the effects of water pollution, and a series of full-page ads in the *Toronto Telegram*, one of which offered a brimming glass of brown, viscous water from the Don River as a refreshment to politicians. Their message reiterated what long-established groups such as the Toronto Field Naturalists had been saying for years: the Don had the potential to be a vibrant green corridor in the heart of the city, a refuge for wildlife and a much-needed space for recreation, and it was worthy of protection. Unlike earlier groups, however, Pollution Probe spoke for a new generation that refused to accept the degradation of the environment as an inevitable consequence of development.

The 1969 funeral was followed by a brief surge of interest in the Don, and a 1971 campaign by the Ontario Water Resources Commission to reduce phosphates in Ontario waterways was successful in raising oxygen levels in the Don and improving aquatic habitat.[5,6] In the summer of the same year, college students hired for the Don Patrol, a joint initiative of the MTRCA and General Foods Ltd., removed more than 200 tonnes of litter from the river and surrounding valley. It wasn't until the late 1980s, however, that heightened public concern for the environment generated new and sustained visions for a restored river environment. On February 23, 1989, responding to concerns from local residents' associations, Toronto City Council endorsed a recommendation 'that the Don River and its related recreation and wildlife areas be made fully usable, accessible and safe for the people of Toronto no later than the year 2001.'

That same year would prove to be a landmark year in the history of the Don. In the spring, the *Globe and Mail*'s *Toronto Magazine* published 'Rebirth of a River,' an article that looked to

5 Paul Theil Associates Ltd., *Strategy for Improvement of Don River Water Quality: Summary Report* (Toronto: Queen's Printer, 1989), p. 4.

6 Phosphates promote excessive growth in algae and other aquatic plants, creating a deadly environment for fish and shellfish by depleting available oxygen and disrupting ecosystem function. This process is called eutrophication.

other cities, including Cleveland and London, for examples of the types of effort and investment required to rehabilitate the Don. Through a series of interviews with Don River advocates from different backgrounds – concerned residents, naturalists, scientists and politicians – author Pat Ohlendorf-Moffat outlined, in broad brushstrokes, a vision for a revitalized river. Significantly, she stressed the vital role individuals could play in regenerating the Don by lobbying municipalities to purchase sensitive headwater lands from developers, participating in cleanup and restoration events, reducing individual contributions to toxic runoff by finding alternatives to pesticides and herbicides, and lessening the use of sidewalk salt in winter. The article was followed by a day-long public forum on the future of the Don at the Ontario Science Centre. Attended by about 500 people, the forum represented a watershed in public awareness about the Don.

### MARK WILSON AND THE TASK FORCE TO BRING BACK THE DON

For Mark Wilson, the Science Centre forum was the beginning of over fifteen years of involvement and leadership in community-based advocacy and watershed restoration. 'That was where the Don caught my imagination and my heart. I learned about the water cycle and how stormwater was polluting the Don. Helen Juhola [of the Toronto Field Naturalists] talked about the great natural habitats and the urgent need for action to preserve them. [Landscape architect] Glenn Harrington told us about the moral imperative to restore the Don so salmon could once again swim and spawn. I learned about what other cities such as Cleveland were doing to restore their trashed urban rivers . . . When councillor Jack Layton stood up and said there were a group of people meeting at City Hall who were going to do something about the Don, I had to join.'

Wilson joined a dedicated group of citizens, City councillors and staff that worked together to develop a proposal for a public task force on the Don River. Completed in May 1989, the proposal presented a vision for a clean, green and accessible Don. The Task Force to Bring Back the Don was created several months later with staff support and a starting budget of $170,000 provided by the City of Toronto and the Toronto Harbour Commission. This unique formula of strong commitment from citizens, coupled with support from the City, has been the key to its success. That, and the passion and energy of its leadership.

Indeed, it is impossible to tell the story of the task force without referring to the leadership of Mark Wilson. Chair of the task force from its establishment until 1998, Wilson was to the citizens' movement to restore the Don in the 1990s what Sauriol was to Don Valley conservation efforts in the 1940s and '50s. In collaboration with other task force members, he built relationships with City officials and federal and provincial government agencies, secured funding from local foundations, corporations and federal granting agencies, attracted citizen support and injected the task force with a flair for playful, innovative communications. On Earth Day in 1990, for example, the task force put banners and wishing wells on the bridges over the Don Valley and asked Torontonians to make a wish for the Don. Recognizing the strategic importance of situating the Don within the broader urban sustainability movement, Wilson and the task force presented a vision of a restored Don River as a living demonstration of this approach in *Bringing Back the Don*, their first report to City Council. The report outlined six key objectives in its restoration strategy for the Don: enhancement of the river mouth; creation of aquatic habitats, including wetlands; restoration of terrestrial habitats; encouraging appropriate uses of the valley; improving access to the valley; and coordinating planning policy for the valley.

From the beginning, the task force has focused on small restoration initiatives rather than larger capital-intensive projects, on shovels in the dirt rather than lengthy studies. Since 1989, task force volunteers – over 10,000 at last count – have planted tens of thousands of trees, shrubs and wildflowers in the Lower Don Valley, and removed many tons of garbage and debris from the West and Lower Don. Forty restoration projects have been initiated throughout the central and lower valley. Visitors to the Don are likely most familiar with the restoration work at Chester Springs Marsh in the flats just south of the Bloor Street Viaduct. Completed in 1996, the marsh provides critical habitat for wetland wildlife. Monitoring between 1996 and 2004 showed that if you build it they will come: wetland species such as the painted turtle and great blue heron, and cattails and other aquatic plants have increased in number and diversity since the marsh was created.[7] The benefits of this wetland habitat have also been felt in the form of natural flood control and water purification: porous soils absorb excess stormwater in times of flood, and aquatic plants improve water quality by filtering

7 While few species of concern (the 'indicator' species of ecosystem health) have been recorded at Chester Springs Marsh, overall numbers of nesting marsh birds are equal to or higher than the average for the Great Lakes region. The number and diversity of amphibians, however, remains lower than the Great Lakes average, according to the Marsh Monitoring Program's *Marsh Bird and Amphibian Communities in the Toronto and Region AOC, 1995–2002* (www.bsc-eoc.org/download/MMP-AOC%20Toronto%20and%20Region.pdf). Native trees and shrubs such as tamarack and red osier dogwood have been especially successful reclaimers of the marsh. Although problems with invasive non-native vegetation persist in the upper dry areas of the marsh, vegetation surveys like Steve Gillis's 2003 *Chester Springs Marsh East Community Stewardship Report* indicate native wetland plants such as arrowhead, smartweed and water lilies now dominate the marsh shoreline.

*Youth volunteer planting event at Beechwood Wetland, 2005.*

pollutants. Finally, the marsh has created a connection with the past, becoming a living reminder of the historic marshlands and swimming holes that could be found in the area in the late nineteenth century.

Attempts to improve public access to the Don have perhaps been even more important than these restoration initiatives, however. Speaking about the lower river in 1989, Wilson recalls that 'there was no public access [to the river] between Pottery Road and the Lake – no recreational trail . . . and no regeneration projects.' Only a public who knew the Don, the task force realized, would take efforts to restore it. The City responded by opening the Lower Don Recreational Trail in 1991 and constructing stairs into the valley from Queen Street and the Riverdale Park footbridge. These access points have played an enormous role in reinserting the Don into the collective consciousness of Toronto residents. Through these initiatives and many others over the last twenty years, the task force has become a model for community-driven environmental restoration projects within an urban context.

### NEW GROUPS AND NEW APPROACHES

Since the creation of the Task Force to Bring Back the Don in 1989, other groups have formed to address environmental concerns in the wider Don River watershed, as Task Force activities have been limited to the pre-amalgamation City of Toronto boundaries. In 1992, the MTRCA established the Don Watershed Task Force and charged it with the mandate to develop an ecosystem regeneration plan for the entire Don watershed. Comprising twenty-five representatives from

*Chester Springs Marsh.*

watershed communities, municipalities, community-based groups and external agencies, the Watershed Task Force completed its report, *Forty Steps to a New Don*, in 1994, and the MTRCA began implementing its forty distinct recommendations immediately. Three general principles guided the Watershed Task Force's recommendations: protect what is healthy, such as clean water sources and habitat linkages in the Don; regenerate what is degraded, including water quality, wildlife habitat and cultural heritage; and take responsibility for the Don by facilitating public access and co-operation across government jurisdictions and between community organizations. The Don Watershed Regeneration Council, a watershed-wide advisory committee, was established by the MTRCA the following year to implement the recommendations of the Watershed Task Force report and monitor its results. Four report cards have since been published describing DWRC progress on regeneration projects throughout the watershed, including the successful daylighting of Mud Creek and the development of a wetland site in the former Don Valley Brick Works, opened as a City park in 1997.

In 1993, the Friends of the Don East used the task-force model to undertake restoration and public education initiatives in the former Borough of East York. After Toronto's seven municipalities amalgamated in 1998, FODE continued its work in the communities east of the Don, hosting tree-planting days, local park cleanups and workshops on sustainable living practices. They initiated the Taylor Massey Project in 2003 with the goal of improving the water quality and natural heritage of Taylor Massey Creek (an eastern tributary of the Don), creating a recreation trail along the length of the creek and organizing a series of innovative 'reach stewardship groups' to align concerned residents and community groups with specific parts of the watershed. The TMP has since become a separate organization, and action at the sub-watershed level continues through groups like the Richmond Hill Naturalists and the Friends of Glendon Forest.

## ONGOING CONCERNS

Even with all of the progress that's been made in cleaning and greening the Don, much remains to be done. A 2007 Environment Canada report gave the Don a water-quality rating of 34.8 out of 100, making it the most polluted river in Ontario and the third most polluted river in Canada. The bulk of this pollution comes not from industry or single-source polluters but from the everyday activities of urban life – flushing toilets, driving to the grocery store, salting driveways in winter. Runoff from streets and parking lots carries oil, road salt, animal wastes and other harmful substances into the storm sewer system, and from there directly into the Don. Almost 1,200 storm-sewer outfalls dump into the Don and its tributaries, and stormwater makes up over 70 percent of the river's flow. This stormwater not only contributes to poor water quality but also causes flooding and erosion of riverbanks. Each year, in fact, massive silt deposits at the mouth of the Don pose threats to ship traffic and necessitate costly dredging by the Toronto Port Authority.

Even more serious are the effects of combined sanitary and storm sewers in the older parts of the city, like the area around the Lower Don and along Taylor Massey Creek in the east end. Every time the city receives a heavy rainfall, these combined sewers overflow, carrying raw sewage directly into the river. 'Wastewater is the biggest problem facing the Don River today,' current task force chair John Wilson observed in a 2005 newsletter. 'I hate the fact that my family and I contribute to this problem every time it rains. I just don't want to feel guilty about flushing my toilet during a storm.' Implementation of the City's Wet Weather Flow Management Master Plan, which will improve the quality and reduce the quantity of urban runoff entering the river, should help relieve some of this guilt.

Existing pollution makes the river unsafe for people to wade or swim in, but for wildlife, the consequences are life-threatening. According to the TRCA, chloride from road salt is harmful to aquatic wildlife at 240 milligrams per litre of water. Don levels are consistently higher than this, reaching a high of 3,920 milligrams in samples taken between 2002 and 2005. Runoff from streets also raises the water temperature of the river to levels only the most adaptable species can tolerate. White sucker, creek chub, fathead minnow and blacknose dace are among the seven exceptionally tolerant fish species that have adapted to the toxic conditions in the Lower Don.

Even with better water quality, however, the Don faces problems daunting to even the most optimistic of its advocates. In a watershed that was once almost entirely forested, only 7.2 percent of forest cover remains. Almost all of the watershed's original wetlands have been filled or paved over, and 85 percent of its lands have been developed for residential or industrial purposes. As subdivisions and industrial parks have replaced farmland in the upper watershed, pockets of critical wildlife habitat have been lost and porous soils paved over, sending more surface runoff into the streams and less into vital groundwater reserves, compromising the quality of the Don's remaining sources of clean water. And yet, much is in store for the Don, and what seem like insurmountable challenges are being taken up in a series of innovative visions for the future of the river and its place in the city.

## LOOKING AHEAD: RESTORING THE LOWER DON

Big plans for the Don River seem to be everywhere these days, from Evergreen's redevelopment of the Don Valley Brick Works into Canada's first environmental discovery centre to plans to recreate a part of the historic wetland at the mouth of the Don. In March 2008, the TRCA and Waterfront Toronto hosted a public presentation of the work underway to naturalize the mouth of the Don. Under the magnificent chandeliers of St. Lawrence

Hall, the room packed with Don River advocates and interested residents, it was possible for a moment to dream big for the Don, to imagine, instead of a river moving through walls of concrete and abandoned industrial land to dump unceremoniously into the harbour, a river surrounded by vibrant wetlands and walking paths, the vision of a clean, green and accessible Don River so long advocated by the task force.

The plan certainly looks exciting: the Don Mouth Naturalization and Port Lands Flood Protection Project would provide flood protection for lands surrounding the Lower Don and establish a more natural river mouth. Since 2005, the TRCA has been working with a team of consultants on an environmental assessment of the project, and in the spring of 2007 a New York–based landscape architecture firm, Michael Van Valkenburgh Associates, was selected through an international design competition. MVVA's Port Lands Estuary proposal best met the two major objectives of the project, which were, according to the design competition jury report, 'to create naturalized mouth and iconic identity for the Don River, and to deliver a comprehensive plan for addressing the area's ecological, urban design and transportation issues.'

While early visions for the project saw the river running through a naturalized Keating Channel, MVVA proposed leaving the channel intact as an 'industrial artefact' while routing the mouth of the river further south through a naturalized wetland environment. The new river mouth would enjoy greater visibility from other points on Toronto Bay, 'reasserting the presence of the river in the city and allowing it to become a symbol of the Lower Don Lands as a whole.' This design, MVVA argued, would more closely reflect the historic course of the river before it was straightened and channelized in the 1880s. The restored wetland and native forests at the mouth would create much-needed habitat for migratory birds and insects and enhance the existing habitat corridor from the river mouth to its headwaters on the Oak Ridges Moraine. Harnessing the Don's natural sedimentation tendencies is a cornerstone of the MVVA vision: excavated sediment deposits from the mouth of the river would be filtered and treated for contaminants in on-site processing centres, then used to cap polluted land and to create landforms such as hills and flood-protection berms throughout the proposed parkland.

In keeping with Toronto's broader goals for waterfront revitalization, the MVVA design ranks social benefits as highly as ecological ones. An 'urban estuary,' its vision embeds the restored river mouth within a parkland complex that includes recreational fields, walking and cycling trails and shoreline spaces for water-based activities. MVVA proposes to develop four distinct neighbourhoods on either side of Keating Channel and on the north side of the Ship Channel, south of the restored river mouth. Each neighbourhood, MVVA states, 'will have the complete DNA of a vibrant city: a mix of life-cycle housing, commercial, cultural and workspaces, public realms, parkland and access to water.' How and when various aspects of the project will be implemented will be determined through the environmental-assessment process, scheduled to be completed in early 2009. Implementation plans will likely reflect, at least in part, the loose plan forwarded in the MVVA concept design: a six-stage process with channel excavation, soil remediation and sediment capture from the river scheduled to occur in phases one and two, and park construction, infrastructure linkages and neighbourhood construction slated for phases three through six.

As chair of Waterfront Toronto's board of directors, Mark Wilson has been well-placed to shepherd the task force's 1991 vision of a naturalized Don mouth through to a time when fruitful partnerships and a commitment by government funders have made the project both tangible and realizable. Established in 2001 by the federal, provincial and municipal governments to fund and oversee the revitalization of the Toronto waterfront, Waterfront Toronto identified the Don Mouth Naturalization as one of four priority projects in their ten-year business plan. In partnership with the TRCA – the lead proponent of the DMNP – Waterfront Toronto has solicited the participation of stakeholders such as the Task Force to Bring Back the Don and the Toronto Port Authority throughout the planning stages of the project. They have also taken care to incorporate earlier visions for a restored Don River mouth into the planning and design selection process. Of the four viable alternatives for a naturalized river mouth, two of the plans were inspired by task force visions for the Don in the early 1990s. These alternatives were weighed along with a modified version of the MVVA design at public and stakeholder meetings in early 2008. Modifications included filling Keating Channel with clean lake water rather than river water, and the creation of a secondary channel perpendicular to Keating Channel to accommodate floodwaters

FACING PAGE
*MVVA's proposed 'Port Lands Estuary,' 2007.*

and seasonal fluctuations in river flow. The revised MVVA design, with its large area of viable land to be naturalized, its provisions for reuse of site materials and its lower management costs for contaminated soils, emerged from these meetings as the preferred alternative.

Work is currently underway to develop a more detailed conceptual design for the preferred alternative, which will form the basis for the environmental assessment (EA) required by the provincial government before construction can begin. A series of detailed studies on the hydrology, soils and other conditions of the Lower Don conducted through the summer and fall of 2008 will inform this conceptual design. Consultation with the task force and other stakeholders has shaped the development of the design in significant ways: as a result of stakeholder feedback, for example, the TRCA has extended its plans for naturalization beyond the river mouth to incorporate the Don Narrows, the channelized portion of the river south of Gerrard Street. Next steps include the submission and approval of the EA and then the development of a detailed, phased design for the naturalization work. Implementation of the project is forecasted to start in late 2010 at the earliest, with a projected completion date sometime in the 2030s (the river mouth component is expected to be completed sooner, within ten to fifteen years).

While optimism and excitement are the prevailing responses to the project, some skepticism exists about its scale and the likelihood that it will be completed within the targeted time frame and budget parameters. Some feel that the massive scale of the project – redesigning an entire city district as well as a naturalized river mouth – will mean that many of us won't see the project completed within our lifetimes. 'The effort is far too slow,' Mark Mattsen, president of Lake Ontario Waterkeeper, commented in an article in the *Toronto Star* on December 7, 2007. 'It's being put off to another generation. The [naturalization] is "window-dressing."'

Like the City's Wet Weather Flow Management Master Plan, the project will likely take twenty to twenty-five years to complete. Smaller, incremental improvements to the ecological integrity of the Don mouth and narrows might, in some observers' eyes, produce more tangible results in the short term. It is difficult to find comparable terrain for comparison because the DMNP is so much larger in scale and scope than other urban river restoration projects. Improvements made to the Los Angeles

River and the Chicago River, for example, have concentrated on water quality and habitat enhancement, but have not involved major relocations of the river channel or redevelopment of surrounding neighbourhoods. Another area of skepticism concerns the proposed method and expense of cleaning up the contaminated soils left behind by former industrial sites around the mouth of the Don and throughout the Port Lands. Rather than attempt to cleanse contaminated soils in areas slated for residential development, MVVA proposes to leave them in place, purportedly preventing leaching by capping the contaminated soil with captured sediment from the Don River. In areas where extensive excavation will occur, such as the river channel, treating contaminated soils and sediments will be unavoidable.

A component of the EA now underway will weigh the effectiveness of different kinds of soil-remediation strategies, and add detail to the concept of an on-site soil-remediation plant to cleanse and recycle soils for use elsewhere in the development. There is some concern, nevertheless, that the management of contaminated soils will not be effective enough, or will cost more than predicted. Still others feel that the Don River and its mouth, situated as it is on public land, should not be dependent upon privatization and condo sales for its revitalization. Public sector funders hope to receive a 14 percent return on their investment through the sale of residential and commercial units, a plan that some feel places too much emphasis on the marketability of the proposed design plans. Whether the project will continue to move ahead as planned, and whether the funds available will prove sufficient remains to be seen. What is clear is this: not since the infamous Don Improvement Project of the 1880s, which straightened and channelized the lower river to make way for rail and industrial interests, has there been so much excitement and political will to imagine a new future for Canada's most notorious urban river. Perhaps, as Charles Sauriol believed and as Mark Wilson continues to remind us, a river that takes a place of pride in the collective consciousness of Toronto residents, rather than a place of shame, might affect the way we interact with the city we call home.

Jane Schmidt
& Frank Remiz

# High Park waterways: Forward to the past

On a sultry summer day in High Park, a geyser erupted with great force. The plume of water, carrying with it rocks, gravel and sand, gushed fifteen metres into the air. Just as the astonished work crew managed to suppress the geyser, *boom*, off blew the caps workers had just fitted on two artesian wells to prepare for the expansion of two stormwater ponds. When all the wells were finally resealed that day in July 2003, the hillside started to weep. And the water that leaked out was blood-red, as if a wound had opened in the earth.

High Park is not usually prone to sprouting geysers like Yellowstone, but it is under great pressure nonetheless. The park exists largely due to the vision of John George Howard who, in 1873, deeded much of the current parkland to the City of Toronto with the provision that the park remain free to the citizens of Toronto in perpetuity, and that it be left in a natural condition. But nature in the park has been compromised and is under stress. Bold action is required to remedy this situation.

While the visible waterways of High Park have changed radically since Howard's time, the geyser revealed a previously unknown and unspoiled feature of the park. As part of a research project involving nine conservation authorities and four municipalities seeking to understand the hydrology of the Oak Ridge Moraine, a test well was sunk in 2003, with the expectation that it would hit bedrock. But, forty-five metres down, engineers discovered something very different. Bill Snodgrass, witness to the gusher and a senior engineer with Toronto Water, explains: 'What we eventually realized was that we'd discovered the southern section of the great aquifer referred to as the Laurentian River running deep in the bedrock from Georgian Bay.' Although geologists had long been aware of the Laurentian's existence in southern Ontario, no one knew that it ran under High Park or, as the geyser attested, that the aquifer was blocked just before reaching Lake Ontario.

The Laurentian River was formed millions of years ago, long before humans were a twinkle in evolution's eye. Predating the formation of the Great Lakes, its channels were carved in bedrock and then buried and sealed by layers of glacial rubble.

Not a rushing river, it takes its time, seeping as slowly as one centimetre a year through channels filled with rocks and stones.

The Laurentian water that reaches High Park started out many thousands of years ago from its Georgian Bay source, and represents a vast reserve of pristine groundwater. Tests indicate that the water in High Park is pure and perfectly potable, but has a high iron content that is too expensive to filter out for drinking. (A slight salt concentration also prohibits its use for watering plants in the High Park greenhouses.) The iron in the water oxidizes rapidly when exposed to air, turning the water red and staining whatever it touches, like the concrete channel by the stormwater ponds in the park's northeast corner [1]. There is the proof of the Laurentian River in High Park. And it is there that the past meets the present, with future ramifications.

[1] ABOVE *Evidence of the Laurentian River in High Park. Because of its high iron content, the water, when exposed to air, stains the Spring Creek channel red (here, a lighter grey).*

[2] ABOVE, LEFT *High Park Mineral Baths, 1914. The site of the Laurentian River discovery, these baths were filled in shortly after this photo was taken. Bloor Street is under construction in the background.*

The Laurentian aquifer has remained relatively unchanged since its entombment, while immense changes have happened on the surface above. The bison, muskox and giant beavers that roamed on ice a kilometre thick over the land mass 25,000 years ago had no impact on it. Neither did the more recent melting and retreating of the glaciers, which created High Park's watercourses, ponds and ravines. The river flowed on, impervious to the First Nations people who occupied the Toronto region, the European settlers like Howard and the bathers who, a century ago, enjoyed the mineral baths near where the geyser erupted [2]. The Laurentian's waters have been immune to the effects of increased human population, built structures, hardened surfaces, creek burials, toxic

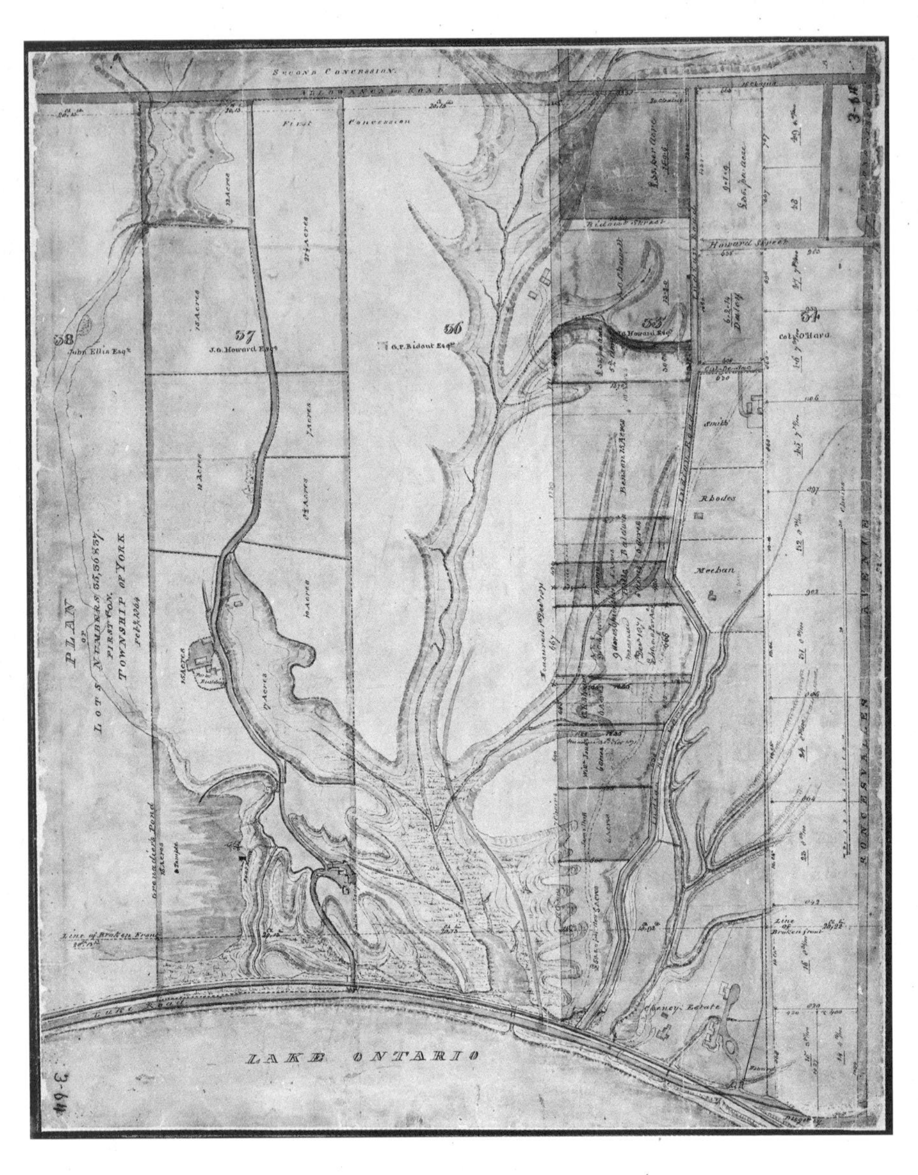


3 *Plan of survey by City surveyor John G. Howard that shows some of the former watercourses in High Park in 1864. Only the east bank of Grenadier Pond appears on this plan.*

stormwater runoff, holes in the ozone layer and climate change – at least, so far.

The surface streams, ponds and wetlands of High Park have not been so fortunate. Much of what has happened to the park's waterways is common to the rest of Toronto: wetlands filled in; rivers and streams diverted, buried and converted into stormwater channels; and direct aquatic connections to Lake

Ontario cut off by transportation corridors and lakefill. Survey maps drawn by Howard – who, in 1837, built the still-standing Colborne Lodge that overlooks the lake – show that water features once comprised almost half of present-day High Park [3]. Today, only 20 percent of them remain.

A significant park feature that has survived from before Howard's time is a rare ecosystem, the black oak savannah. This open woodland seems the antithesis of water because its typical species survive on dry soil and are well-adapted to drought and ground fires. But the conditions were created by water: the sandy soil is the result of deposits laid down in ancient Lake Iroquois about 12,500 years ago. In 1989, the Province designated High Park's savannah as an Area of Natural and Scientific Interest worthy of protection.

High Park is also unusual in that it drains two small watersheds, Spring Creek and Wendigo Creek, whose long-buried headwaters lie north of Bloor Street. Winding, deep-set roads like Clendenan and Glendonwynne follow the original route of Wendigo, just as Medland and Glenlake streets trace Spring Creek's former course. Both creeks emerge into the open in High Park, though neither is a natural watercourse any longer. Little more than stormwater channels, the creeks are contained by reinforced edges to accept the sudden rush of rainwater from catchment areas north of the park, now largely covered by hard surfaces.

4 TOP *Restoration work at Lower Duck Pond. The string grids prevented birds from feeding on the new aquatic plants until they were established.*

5 *Skaters on Grenadier Pond, c. 1910.*

Much of the recent waterway-restoration work in High Park has been for stormwater management. Settling ponds were constructed at the north end of the two creeks to reduce the sediments, nutrients, heavy metals, oils, animal feces and bacteria that pollute the stormwater entering the park. At the south end of the park, Wendigo opens into Grenadier Pond, while Spring empties into Lower Duck Pond. Both ponds have been extensively planted in recent years with cattails and other wetland plants to filter pollutants and provide wildlife habitat [4].

Grenadier, Toronto's largest natural pond, has long been a favourite recreation area. Ice skating was once a popular winter activity there [5]. (It is now forbidden.) In the summer, people often birdwatch, fish or stroll by the pond. Even royalty has visited: in 1959, Queen Elizabeth opened the Maple Leaf Floral Display beside the pond. Concrete edges were added to the pond to make a neat promenade for Her Highness, much to

6 7 *The Grenadier Pond shoreline, before and after 1995. The concrete curb was removed and replaced with native wetland plants.*

the detriment of the pond's fish. The concrete has since been removed and the edges softened so fish once again have a place to shelter and lay eggs [6 7]. The addition of shoal fingers and log tangles has improved fish habitat throughout the pond, which is stocked with largemouth bass, a top-level predator that helps control the imbalance in the number of smaller species. After a pollution-related population decline in the 1990s, a dozen fish species are now making a comeback.

The north end of Grenadier, home to the lovely yellow iris and the regionally rare sweet flag, has been restored with a variety of aquatic plants and is now a virtual marsh frequented by wading birds such as great blue herons, night herons and egrets. During spring and fall migration, over twenty different waterfowl species can be seen stopping over in Grenadier.

All of High Park's waters flow to Lake Ontario – eventually. Both streams drain into Grenadier Pond (by pipe, in the case of Spring Creek) and are then diverted by underground conduits to the Humber River. However, because of the Laurentian River running under it, High Park has a more natural water connection to Georgian Bay 200 kilometres to the north than it does to Lake Ontario half a kilometre to the south.

That makes life complicated for aquatic mammals. At dusk in the summer, a lone beaver has occasionally been spotted gliding though Lower Duck Pond. But how did it get there? Between the pond and Lake Ontario, there are ten lanes of roadway, two sets of streetcar tracks and four sets of train tracks. And there aren't any direct water connections to the east, west or north either.

Ralph Tonninger, from the TRCA's Restoration Division, postulates that 'the beaver was probably a young male sent off from his family lodge in the Humber region. Beavers are not just aquatic; they travel very well by land.' Highly attuned to the sound of running water, be it stream or storm sewer, he would have alternately walked and swum from one water source to another. Just like Burt Lancaster in the 1968 film *The Swimmer*, who swam home from one backyard pool to another. For Lancaster's character, it was existential angst; for the beaver, it's a necessity. There may be another explanation. 'Beavers are nocturnal. Maybe he just walked from the lake, when no one was looking,' said Tonninger. A forlorn notion: one beaver waiting for a gap in traffic.

The challenges for aquatic wildlife are compounded by increasing development pressures all around High Park. A controversial condominium was recently built at the northwest corner of the park, despite fears of the waterways being adversely affected. In the last few years, dozens of condos have sprung up along the Queensway to the west, and more construction is underway closer to the park. Mature neighbourhoods to the north, east and west have also seen similar development as a result of calls for high-density living. Along Bloor Street, the park's northern border, several dozen houses and buildings are boarded up and falling apart, likely to be replaced with more condos. The inevitable reduction of green space and additional hardened surfaces will increase wet-weather-flow volumes, and the added population could intensify the pollutants entering Wendigo and Spring creeks.

There are pressures inside the park as well. High Park is a multi-use area with many different and sometimes conflicting activities. In the 1960s, authorities cut down a number of trees to make way for playing fields. Dogs off leash and mountain bikes trampling natural areas go contrary to restoration principles. Irresponsible fishing practices in Grenadier Pond have resulted in abandoned fishing lines and hooks wrapped around duckbills and lines being cast out into bird nests. Accommodating all demands while maintaining ecological objectives is becoming increasingly difficult.

Countering these negative trends requires citizen action with a water focus. Sometimes making an effort to restore the natural environment is like taking two steps forward and one step back. 'Water presents an even greater challenge than land,' says Gary Wilkins, the TRCA's Humber Watershed Specialist. 'What ends up in water is out of sight, out of mind.' Creating a greater awareness of the aquatic environment would help propel positive change. There should be a Water Stewardship Program in High Park, just as there is a Volunteer Stewardship Program dedicated to restoring the black oak savannah. This WSP could advocate for green building practices and improved infrastructure, and educate the public about reducing the impact of stormwater in the catchment areas affecting High Park. It could also work to protect groundwater, naturalize the streams and increase the amount of wetlands in the park, with the aim of increasing biodiversity and improving fish and wildlife habitat in the park's waterways. The WSP would have an active role on the High Park Community Advisory Council, a volunteer-based organization that provides the City with input on how to protect the park for future generations.

Restoring the waterways of High Park also means making some bigger connections. The City's Western Waterfront Master Plan, still in the early stages of public consultation, has proposed cutting off Colborne Lodge Road to traffic and turning it into a pedestrian link between the park and the lake. One step more and it could become a green corridor; a running leap and it could go even further. It could become a blue corridor, Spring Creek would again flow into Lake Ontario, and Grenadier Pond would extend to the lake in a marshy expanse. Below Colborne Lodge, the sound of lapping waves would replace the roar of traffic. Thickson's Woods in Whitby is surrounded by mammoth factories, yet its direct connection to Lake Ontario is a marshy magnet for birds looking for green space and sheltered water. High Park has even greater potential.

Known as the jewel of Toronto parks, High Park could become an example of a thriving nature preserve in the urban fabric. For that to happen, it must have one sole function, and must stop trying to be all things to all people. That means eliminating the 'urban' and 'recreation' from the park. There is already a movement to ban cars. The City could take it further and close the park to bikes and dogs. Some very charming features, like the cultivated gardens and playing fields, would have to be relocated. The park would be a place to walk, not on hard surfaces, but on earthen trails, save for the occasional unobtrusive bridge. Visitors could meander through savannahs, meadows and woodlands, over running streams and out into marshlands. The only buildings, other than a few inconspicuous washrooms and historic Colborne Lodge, would be used for nature education. Spring and Wendigo creeks would run as natural steams into the lake via marshes and wetlands, less and less burdened by stormwater pollution and erosion. Salmon and pike could once again spawn in Grenadier Pond, which in the spring would resound with the mating songs of frogs and toads. Much like a provincial or national park, High Park would become a true wilderness, but in an urban area. Tommy Thompson Park, created from lakefill since 1959, is setting a local precedent for urban wilderness: it is going in exactly this direction, and the High Park ecosystem outlives it by over 10,000 years! It would take a lot of work to restore these aquatic and terrestrial environments, but eventually the ecosystem could take care of itself.

Declaring High Park a wilderness park would not make it elitist; on the contrary, it would welcome everyone to enjoy the park with one purpose in mind. The connection to nature is missing in cities; there's now even talk of a nature-deficit disorder among children. Reconnecting to the natural environment makes for healthy individuals and a healthy society. There would be resistance from those who currently enjoy the park's recreational features, but future generations would applaud the foresight that created a wilderness heritage in Toronto.

Moving forward by restoring High Park to its past incarnation is a radical idea, just as it was in 1909, when the Toronto Guild of Civic Art told a skeptical City Council that Toronto's natural landscape gave the city its identity and was worth protecting. The City is now less disbelieving. Plans and studies flow out of City Hall expounding sound principles and strategies for ecological restoration and the reduction of stormwater volumes and pollutants. But there are serious financial constraints and development pressures that make it unlikely these plans will come to fruition. It will take dedicated citizens – well-informed and willing water stewards – to protect and bring back High Park's natural environment. Restoration will be a slow process, but there will be visible positive changes occurring each year. In time, beavers will merely swim the short, direct route from Lake Ontario to High Park's ponds.

The discovery of the Laurentian River started with a bang; High Park's story should not end with a whimper of degraded nature or development opportunity. Instead, it should continue as a quiet and contemplative nature preserve. To be still is to discover what was always there but unnoticed in haste. There is much to be said for not rushing like the fury of stormwater, unable to be absorbed, instead adopting a slower pace, steady and indomitable like the Laurentian River, enduring and lasting and wise as High Park could be in the future.

## Bringing in the rain: Is rainwater harvesting the solution to Toronto's energy and water needs?

Located on King Street just west of Parliament, the $30 million SAS Canada building is one of the greenest office buildings in Toronto [1]. Completed in 2005, it was the first commercial building in the city certified under the Leadership in Energy and Environmental Design (LEED) program. The software company's head office has been designed to save energy, reduce waste, create a healthy workspace for the occupants and also conserve water. One of its most unique features is a comprehensive rainwater harvesting system, which collects rainfall from the landmark 'inverted' roof and the above-ground terrace levels around the building. The rainwater is conveyed in pipes to a storage system of four tanks (totalling 40,000 litres) in the lower level of the building, filtered, then distributed to provide water to all toilets and urinals. Treated rainwater is distributed through the building using an energy-conserving pressurized tank system, and the intermittently operating pumps use less energy than conventional pumps to move water through the building.

Given the roof area of the building and the average annual precipitation estimated in downtown Toronto, the building's rainwater harvesting system captures almost 1 million litres of rainfall and snowmelt per year. Jerry T. McDermott, SAS Canada's manager of real estate development, estimates that the system reduces the building's potable water demand by up to 80 percent in a 'wet' year. And stormwater runoff from the site has been almost completely eliminated, reducing the strain on the historic neighbourhood's sewers.

The SAS building is a prime example of how companies and individuals can reconsider the value of urban spaces and use resources more wisely. Filling the building's toilets with rainwater is not only water-wise, it's also reducing the amount of energy the City uses to treat and pump potable water and wastewater.

While many people recognize the importance of water conservation, few realize that water conservation and energy conservation go hand in hand. Running the water-supply system accounts for a staggering third of Toronto's municipal electricity bill, or over $47 million in electricity costs annually. In 2008, RiverSides, a Toronto-based non-profit

**1** SAS *Canada's King Street East headquarters.*

corporation whose mission is to protect rivers and lakes by eliminating stormwater runoff pollution, undertook a study for the Toronto Atmospheric Fund (TAF) on the energy costs of municipal water supply and the conservation potential of rainwater harvesting. RiverSides' research shows that rainwater harvesting, as a water-supply alternative, can offer the City significant water and energy savings and can reduce associated greenhouse-gas emissions.

So, what is rainwater harvesting? Just as it sounds, rainwater harvesting is the collection of rainwater for use as water supply. A sophisticated system can supply an entire building's water needs, including drinking water, while a simpler system can take the form of a rain barrel whose catchings are used to water a backyard garden. Whereas rainwater harvesting is used as an essential water source in cities around the world – it is the only source of water in many arid regions, and is required in new buildings in Germany, parts of Australia, China and India – it remains relatively rare in Canada.

We can blame the widespread availability of inexpensive fresh water and the low cost of energy for the slow uptake of rainwater harvesting in Canada. In Ontario, dual plumbing

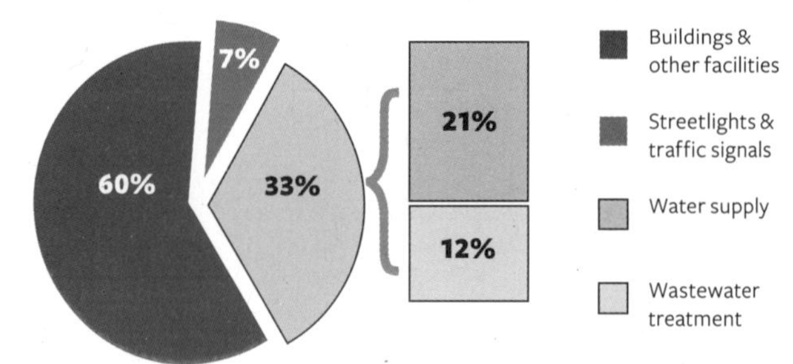


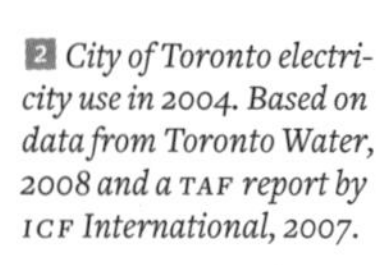

*City of Toronto electricity use in 2004. Based on data from Toronto Water, 2008 and a TAF report by ICF International, 2007.*

systems that enable rainwater harvesting were made permissible only in 2006 by an Ontario Building Code amendment. Prior to this, it was practiced and permitted only on a case-by-case basis. There is now a small but growing number of buildings that incorporate sophisticated harvesting systems for indoor and outdoor water supply in Toronto, including the Healthy House on Sparkhall Avenue (which was the winning design in the Canada Mortgage and Housing Corporation's 1996 Healthy Housing™ Design Competition), the aforementioned SAS Canada building, the Minto Midtown condominium on Roehampton Avenue and a pilot project at Exhibition Place. Demand has been slow to grow, though, and there is still an overwhelming lack of public interest in rainwater harvesting and political will to push the agenda forward.

In the meantime, our water system is affecting much more than just our water resources. A 2007 TAF report, based on 2004 data, found that Toronto Water accounts for 33 percent of the total electricity and 6 percent of the total natural gas used by the City of Toronto.[1, 2] Toronto Water is also responsible for 10 percent of the City's total greenhouse-gas emissions, a total that includes the emissions from the City's vehicle fleet and the methane released from its landfills.[3] Measured in carbon dioxide equivalents, or $eCO_2$, the greenhouse-gas emissions produced by Toronto Water operations add up to a massive 159,315 tonnes of $eCO_2$ per year [2].

The most astounding thing about this electricity use isn't that it cost Toronto Water over $47 million in 2006,[4] but that much of that energy was wasted, since the vast majority of water treated to drinking-quality standards is not used for drinking. Fifty-one percent of water in Toronto is used in residential

1 ICF International, *Greenhouse Gases and Air Pollutants in the City of Toronto: Towards a Harmonized Strategy for Reducing Emissions*, June 2007. Prepared for the Toronto Atmospheric Fund and the Toronto Environment Office. Greenhouse gases are calculated using Environment Canada's emission factor for the Ontario energy supply mix, minus 10 percent for transmission losses. In 2004, the coefficient was 244 g $eCO_2$/kwh.

2 The average Canadian municipality uses between 30 and 60 percent of its electricity on water and wastewater treatment. Larger cities tend to spend a smaller share as they have higher energy bills for social housing and more facilities and services. Power Application Group Inc., *Ontario Municipalities: An Electricity Profile*, January 2008. Prepared for the Independent Electricity System Operator.

3 Methane release from landfills is the largest source of greenhouse-gas emissions, at 45 percent of the total municipal emissions.

4 Toronto Water, 2008.

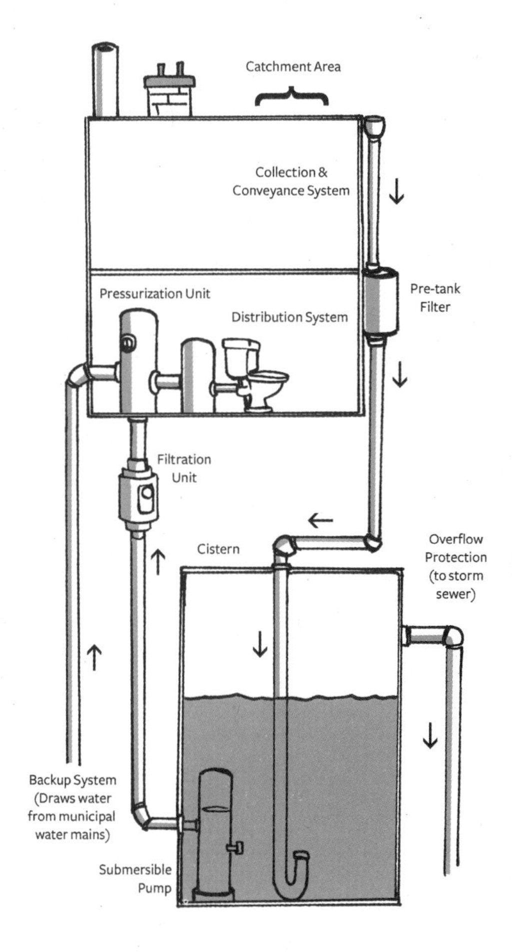


3 *A typical rainwater harvesting system plumbed into a building includes these basic components: a tank, pre- and post-tank filters and pumps, backflow-preventers (to prevent contamination of the municipal supply), overflow outlets and a pressurization unit.*

buildings but, on average, only 8 to 10 percent of the potable water in a typical household is used for drinking and cooking. Waste also occurs in the combined sewer system, where rainwater is combined with sanitary sewage and then unnecessarily pumped to wastewater treatment plants for processing.

Rainwater harvesting is a simple technology that offers a solution to many of Toronto's water issues [3]. Harvested rainwater with minimal filtration can be used inside and outside buildings for non-potable purposes like landscape irrigation and toilet flushing, saving valuable potable water and the energy used to treat and move that water. Rainwater harvesting also can reduce the major cause of urban water pollution: stormwater runoff. Fast-moving stormwater picks up heat and contaminants from roads and rooftops and carries

[4] *Rooftops of the* CBC *Broadcast Centre and other buildings, looking south from Metro Hall,* 2008.

them directly to our rivers and lake, causing the erosion of riverbanks, stirring up sediment, destroying cold-water fish habitats and poisoning aquatic wildlife with road salts, oil, gas, pesticides and fertilizers. Unmanaged stormwater is also the major cause of road washouts and basement floods. Keeping rainwater out of storm and combined sewers reduces the strain on the wastewater treatment system and reduces combined sewer overflows.

A 2005 Green Roofs Study conducted by Ryerson University for the City of Toronto determined that the roof surface of Toronto's large commercial, institutional, industrial and multi-unit residential buildings adds up to about 134 million square metres [4].[5] By multiplying that surface area by 792.7 millimetres (the average annual rainfall in Toronto, according to Environment Canada), and assuming a 20 percent loss for evaporation and inefficiency, the estimated potential captured rainfall is 85 billion litres, or 85,000 megalitres (ML). This is equivalent to nearly 17 percent of the City's annual water supply and translates into a potential savings of 57 million kilowatt hours of energy, 11,300 tonnes of $eCO_2$ and $5 million per year. Averaged over the year, it's also equal to a city-wide savings of 232 megalitres of water per

5 The study identified buildings with flat roofs over a minimum size of 350 square metres, which are assumed to be commercial, industrial or institutional. (Ryerson University, *Report on the Environmental Benefits and Costs of Green Roof Technology for the City of Toronto*, October 2005)

[5] Table A

Rainwater harvesting potential and water-supply energy savings (based on 2006 data from Toronto Water published in 2008)

| | |
|---|---|
| City of Toronto roof surface | 134,000,000 $m^2$ |
| Toronto average annual rainfall* | 792.7 mm |
| Expected volume of rain captured (80% capture) | 84,977 ML |
| Energy savings | 57,189,521 kWh |
| Cost savings | $5,013,643 |
| Greenhouse-gas emission reduction** | 11,302 tonnes $eCO_2$ |

* Precipitation data obtained from Environment Canada reflects the average annual precipitation including snowfall.

** Greenhouse-gas emissions were calculated using the 2006 coefficient of 198 g $eCO_2$/kWh for energy mix for Ontario.

[6] Table B

Rainwater harvesting potential and wastewater treatment energy savings (based on 2006 data from Toronto Water published in 2008)

| | |
|---|---|
| Total roof area for the City of Toronto | 134,000,000 $m^2$ |
| Toronto's core roof area (40% of total) | 53,600,000 $m^2$ |
| Core roof area connected to combined sewers (80% of total) | 42,880,000 $m^2$ |
| Annual rainfall | 792.7 mm |
| Expected volume of stormwater diverted from combined sewers and wastewater treatment* | 34,000 ML |
| Energy savings | 14.6 million kWh/year |
| Cost savings | $1.1 million/year |
| Greenhouse-gas emission reduction** | 2,910 tonnes/year $eCO_2$ |

* For simplicity, it is assumed that all stormwater runoff from a building with downspouts connected to the combined sewer system enters the system and is treated. In reality, some runoff is discharged without treatment in major storms and overflow events.

** Greenhouse-gas emissions were calculated using 2006 coefficient of 198 g $eCO_2$/kWh for energy mix for Ontario.

day – significantly higher than the City's Water Efficiency Plan target savings of 150 megalitres/day [5].[6]

In addition, according to Toronto Water, approximately 80 percent of buildings have eavestroughs and downspout systems that connect to combined sewers. The number of buildings in the combined-sewer area of central Toronto represents

6 *City of Toronto, Water Efficiency Plan*, 2002.

roughly 40 percent of the city's total number of buildings [6].[7] Assuming a correlation to the city's total roof area, that means there are approximately 53 million square metres of roof area in the combined-sewer area. If you multiply the average annual rainfall by the estimated 80 percent of core roof area that's connected to combined sewers, you get 34,000 megalitres of stormwater that could be diverted from these sewers, or nearly 8 percent of all wastewater treated by the City annually. This diversion also has the potential to save nearly 15 million kWh of energy and 2,910 tonnes of $eCO_2$ expelled annually as a result of wastewater pumping and treatment.

Add all these energy savings up and you get 72 million kWh, or 13 percent of the total electricity used by Toronto Water. Moreover, implementing wide-scale rainwater harvesting would also prevent 14,212 tonnes of greenhouse-gas emissions from being released into the atmosphere, and could save the City $6.1 million in operational electricity costs.

Most important to the long-term well-being of the city, though, is that reducing the demand for energy and water would also lessen the pressure on existing infrastructure, and would cut the need to expand the system to accommodate a growing population. Using Toronto's Water Efficiency Plan calculations from 2002, and assuming a 30 percent increase in construction costs, a permanent reduction of one litre of daily water demand would save the City $0.61, and a one-litre reduction in the daily wastewater volume would save $0.85 in one-time infrastructure expansion costs.[8] Using these calculations, if rainwater harvesting could reduce daily water demand by 232 megalitres, Toronto Water has the potential to save up to $141 million in annual infrastructure costs.

At present, the City bases the size of its water supply and treatment system on peak demand levels. Thus, any reductions in the peak demand periods are especially important. Captured rainwater can be a source of water during peak periods, as the increased demand – sometimes up by two-thirds – is primarily for non-drinking purposes like irrigation and cooling. The additional energy required to process water in a high-peak period lasting five days (a typical summer heat wave) can amount to almost 3 million extra kilowatt hours, generating almost 560 additional tonnes of $eCO_2$. Not surprisingly, peak water demand in the summer months coincides with peak energy demand, increasing the overall strain on

7 Data provided by City Planning Division, 2006.

8 City of Toronto, Water Efficiency Plan, 2002. Based on various Statistics Canada construction indices, we estimate that construction costs have increased roughly 30 percent from 2002 to 2007.

the power grid, the risk of blackouts and the risk of poor air quality due to the use of additional power sources, including 'dirtier' sources such as coal.

In spite of the benefits it could bring to the municipality and the community at large, buildings with rainwater-harvesting systems have been slow to materialize. That's mainly because the long payback period and other barriers make it unattractive to developers and building owners. Costs for systems vary depending on the size of the cistern, its location (i.e., if excavation is needed), the equipment needed (i.e., filters, pumps, piping) and whether the system is being installed in a new building or in a building that's being retrofitted.

The rule of thumb for calculating the cost of the system is an average of $1 per litre of capacity, though variables such as roof size, water demand, cistern size and precipitation patterns all affect the conservation potential, making it an unpredictable method to conserve water and money. However, a building with a plumbed-in system that supplies only non-potable needs can expect to fulfill anywhere from 10 to 90 percent of its water needs.

Over the past few years, Toronto has instituted policies in support of rainwater harvesting, particularly in the Toronto Green Development Standard, which contains performance targets and guidelines set out by the City that promote environmentally sustainable development. Rainwater harvesting is also promoted as a key best-management practice for reducing stormwater in Toronto's Wet Weather Flow Management Master Plan, and it is generally supported by various other green building policies. However, improved policy implementation would see rainwater harvesting pushed to the next level, beyond the fringe of green building and into a mainstream water solution.

Toronto's award-winning Green Roofs program is a prime example of how the City can encourage sustainable building technology in a creative way. In 2001, it implemented a pilot study of different types of green roofs in small plots at City Hall and then launched a two-year pilot subsidy program accessible by any building in the city. The City is planning to launch a larger Eco-Roofs Incentive Program in 2009 that will incorporate green roofs and 'cool roofs' (reflective roofs that reduce solar-heat gain).

The Green Roof strategy included public and expert consultation, the production of research reports, the creation of new policies, an information website and a fact sheet for builders, developers and engineers outlining building code and regulatory implications for green-roof development in Toronto. As further evidence of the City's commitment to green-roof education and research, Toronto will host the 2009 World Green Roof Congress. Using a similar model for rainwater harvesting would increase public acceptance, reduce the development approvals time for building owners and help them to take advantage of the rain already falling on their roofs.

A 2006 study by the marketing consulting firm Freeman Associates showed that building and property owners consider the development approvals process long and cumbersome, and believe they could be slowed further by using innovative building techniques. The additional costs, most importantly in terms of time needed for design and the approvals process, are considered more significant than the system costs itself.[9] Clear information on approvals for building owners, designers, engineers and the municipal staff and building inspectors who review the applications could help to make rainwater harvesting more feasible. The City-created Green Roofs brochure is the ideal information tool for all parties. Demonstration projects could also provide the needed examples and technical information, as well as help to build capacity in the construction and design sector.

The savings for the municipality generated by industrial, commercial and residential rainwater harvesting could also be used to create financial incentives for building owners, either through programs like Eco-Roofs or as rebates tied to a water-conservation target, as is done for water-efficient toilets. Currently, industrial, commercial and institutional buildings are eligible for a small subsidy through Toronto Water's Water Buy Back Program, which gives commercial building owners a one-time grant of $0.30 for every litre of permanent reduction in daily demand upon completion of water-conservation renovations like the installation of low-flush toilets. A retrofitted rainwater harvesting system could be eligible for this program if the water is used inside the building.

At the present time, the TRCA's Sustainable Technologies Evaluation Program is monitoring the actual-versus-projected water savings and operations of three demonstration projects: the Metro Label Company's printing facility (another LEED building) in Scarborough, the Minto Midtown condo towers

9 Freeman Associates, *Action Plan for Sustainable Practices: Implementation Strategies for the Residential and Business Sectors in the Greater Toronto Area*, November 2006. Prepared for the TRCA, the City of Toronto, the Region of Peel, the Regional Municipality of Durham, York Region, the CMHC and Environment Canada.

**10** TRCA (under the Sustainable Technologies Evaluation Program), *Performance Evaluation of a Rainwater Harvesting System Toronto, Ontario: An Interim Report*, May 2008.

and Brookside Public School in Scarborough.[10] Data gathered at these sites will provide valuable local information for commercial applications in Toronto.

Further collaboration between building owners, government bodies and the experts who design and maintain the systems is needed. Unlike the green-roof industry, there are few experts and designers in Toronto who create these systems, and virtually no local capacity for creating the specialized pumps, filters and cisterns now largely imported from Europe, Australia and the U.S. Without a specific rain-harvesting industry association or network, there is no one-stop shop for researchers, property owners, service providers, governments or the public to go to learn, share information and lobby together.

Toronto requires massive inputs of energy and water to function. The steady supply of clean and safe water and reliable energy sources is a major concern for the future; it is essential that this supply is managed in an ecologically sustainable way. If we take up the relatively easy challenge of rainwater harvesting, we can lighten our energy and infrastructure costs and reduce the pollution of our waterways and atmosphere. The ripple effects of rain harvesting are positive and travel toward many different goals simultaneously. We've been letting this resource slip through our fingers far too long; it's far more than just money down the drain.

Lorraine
Johnson

# Bogged down: Water-wise gardeners get the flush

It was an exhilarating, terrifying storm – the kind where you watch the sky, *oohh*ing and *aahh*ing with each flash and crack, wondering how the cats are coping with the noise, rain and wind. My friends and I sat on the porch as the cold front moved in, bringing with it a torrential downpour. Across the city, thousands of basements were flooding, ravine banks were collapsing and low-lying roads were being swamped with water, dead fish and broken tree limbs. As we chatted, occasionally cowering at a particularly violent burst, we raised our glasses to salute the force and power of this rain. The storm felt cleansing, and so we gave little thought to the rivers forming on the street or to the pollution being picked up from roads and driveways and lawns and swept into sewers. If we had turned our attention to this soupy mix of old oil, animal waste and fertilizer runoff, we could easily have predicted exactly what happened: Toronto's aging and overloaded sewers released the storm's contaminated load directly into rivers and the lake.

On Mulgrove Drive in Etobicoke, just a few kilometres away from where we sat watching the storm on that spring night in 2000, rivers of water gushed into stormwater ditches – shallow, grassy depressions that run alongside the road. Although grass is permeable, it is hardly better than pavement at slowing the flow of fast-moving runoff. So, during heavy storms, water on Mulgrove Drive runs along the ditches and into the storm sewers, and is transported to Renforth Creek and then into Etobicoke Creek, completing its journey south to Lake Ontario near the beaches at Marie Curtis Park. Whatever gunk it has picked up along the way is dumped into the lake. It's no wonder the beach at this park is regularly closed to swimming in the summer.

One of those stormwater ditches on Mulgrove Drive stands out, though. Planted with native wet-meadow species instead of regular turf grass, and boasting more than forty different varieties of wildflowers, sedges and native grasses such as columbine, fox sedge and little bluestem, the ditch garden slows the flow of water and encourages it to seep into the ground. This act of biomimicry – defined as design modelled on nature – is the work of Douglas Counter, the gardener who lives at 52

Mulgrove Drive and who created this planting – an extension of his front-yard meditation garden – with an explicitly environmental goal in mind. Douglas, a graphic designer whose meticulousness in his professional work is matched by the tough but tender care he lavishes on his garden, remembered his father's stories of swimming at Sunnyside Beach as a child, recognized the connection between storm runoff and polluted waterways and decided to do something about it. He looked to nature, considered the wet meadows that form in low-lying areas and found his inspiration in the moisture-loving native plants that flourish in these natural conditions. Combing nursery catalogues, he chose plants for beauty and function, his designer's eye creating a lush and colourful combination of species to enhance water infiltration. His garden is a small-scale, local gesture toward a cleaner lake. When it rains, there's little runoff. The water goes where it should: deep down into the earth.

A few months after that violent storm in 2000, Counter received a letter from a City of Toronto lawyer that warned that his ditch planting was 'an unauthorized use or occupation of the City-owned road allowance in a way that creates a safety hazard to vehicular and pedestrian traffic.' Counter was given two choices: remove his ditch garden or apply for a permit to plant in the ditch at a cost of $700 plus an annual fee of $630, a permit that any neighbour within sixty-one metres of his property line could oppose. In addition to these conditions, he would also be required to get $1 million in third-party injury and property damage insurance. He was given until December 8, 2000, to remove his garden and replace it with sod.

Counter had no intention of tearing up his garden, however, and it seemed the weather was on his side. On December 7, the first snowfall of the year began at dusk and didn't let up until the next morning. The ditch at 52 Mulgrove was buried under a foot of snow, dried seed heads flattened by the force. This was the beginning of a winter of unprecedented snow cover in the city, of one-metre-high snowbanks along the sides of roads courtesy of City plows, of drivers peering around huge obstructions while backing out of driveways onto slippery roads. It was also the beginning of Counter's legal battles to defend his ditch.

The Ontario Superior Court of Justice is not normally the place where heated debates about field pussytoes, prairie smoke and swamp milkweed take place. And it's rare that a

*Douglas Counter's front-yard and ditch gardens.*

garden is at the centre of a constitutional argument. But in the case of *Counter v. City of Toronto* (2002), Counter's lawyers argued that the City's bylaws violated his right to freedom of conscience, religion and expression as guaranteed under the Canadian Charter of Rights and Freedoms. Picking up on a 1996 case in which Toronto gardener Sandy Bell had successfully challenged the City's Grass and Weeds Bylaw, which arbitrarily limited some plants to a height of eight inches, Counter's lawyers argued that he should have the right to express his environmental values through his ditch garden, even if it was on City property. The court agreed, concluding that 'while it is not the purpose of the by-laws to constrain expression, it is their effect.' Ruling that Counter's ditch garden was protected by the Charter's freedom of expression clause (a decision later upheld by the Ontario Court of Appeal), the court recognized – for the first time in Canada – that citizens have the protected right to express pro-environmental values on public land, subject only to safety considerations. Counter could keep his ditch garden (despite all the dark talk about hazards, no alterations to Counter's planting were required for safety reasons), and other citizens could follow his lead.

The judge was explicit in his written ruling that the City should be helping gardeners like Counter by drafting policies. In it, he wrote that 'it seems to me it has now become critical that the City develop and implement a coherent strategy to deal with natural gardens, which all agree [sic] have become

Douglas Counter's ditch garden.

increasingly popular.' Later in his judgment, he returned to this theme: 'I repeat that the City can and ought to avoid problems of this sort by developing and implementing specific guidelines to deal with the critical issue of natural gardens and their enormous environmental significance.'

Instead of building on this opportunity to encourage small-scale solutions to stormwater runoff, however, the City has remained silent on the subject of ditch infiltration gardens. There are no guidelines as to height limits, no details as to what constitutes a safe design, no standards by which to evaluate compliance. Counter has researched what other cities are doing and points to Seattle as particularly forward-thinking: 'Under their SEA [Street Edge Alternatives] Street program, residential streetscapes have been re-engineered to mimic nature's function. Paved road surfaces have been reduced, and planted swales have been created along road edges to act as stormwater infiltration gardens. Complementing this, the City promotes the conversion of lawns to native gardens.' Counter would like to see something similar in Toronto: 'The City should develop clearly defined and easy-to-follow guidelines that don't infringe on natural gardeners' hard-won rights.' While policies to guide boulevard gardeners are lacking, Counter's own garden continues to flourish. Passersby stop almost daily to see what new flowers are blooming, and Counter no longer worries that City workers will show up with power tools. His garden, for a while

*Deborah Dale's bog garden, pre-razing.*

a private/public battleground, has returned to its quieter days as rain-nourished growing ground.

At the opposite end of the city, in Scarborough, Deborah Dale, a biologist with the quiet, self-deprecating demeanour of someone who prefers the sidelines to the spotlight, has also been engaged in a small-scale solution to the runoff problem, never dreaming that her efforts would draw unwelcome attention. In the late 1990s, she disconnected her downspout from the City sewer (something homeowners were encouraged and are now required to do through the City's Downspout Disconnect program) and created a bog garden at its base. Like Counter's planting, Dale's small bog was a landscape modelled on nature, with water treated as a resource, not a problem. The downspout directed each heavy rainfall's inundation to plants happy for a soaking. Instead of running into the sewers, the water nourished Dale's bog.

Until August 21, 2007, that is, when Dale came home to a scene of total destruction. Her front garden – mere hours ago made up of an array of native species such as butterfly milkweed, vervain, culver's root; shrubs such as New Jersey tea, swamp rose and grey dogwood; and trees such as red oak and pawpaw – had been razed, a decade's worth of effort reduced to shorn stubble. She phoned the police, only to discover that her front garden – the bog garden, the boulevard garden, even trees and shrubs – had been cut down by the City.

*After.*

A couple of days later, Dale was served notice of other alleged infractions in her backyard: weeds, heavy undergrowth and dead branches on the interior of a pine tree (branches on which Dale had hung birdfeeders). She appeared at a Property Standards Committee hearing to defend herself against the allegations and to ask what 'weeds' they were talking about (she was not growing any species on Ontario's Noxious Weeds list), and to find out what was meant by 'heavy undergrowth' (one person's heavy undergrowth might be another person's successful groundcover). The committee discussed each allegation without any further action required by Dale on any of the complaints. However, Dale was told that the committee was going to order a fire inspection of her back garden; there were concerns that the mulch in her garden (mulch used by gardeners everywhere, including the parks department) was a fire hazard. At the close of the hearing, Dale's lawyer served the City with papers initiating a lawsuit.

No court date has been set yet, but there's no doubt that Dale's maintenance practices will be intensely scrutinized should this battle reach the courts. Time will be spent debating 'construction debris' (the inspector's term) versus 'Pennsylvania limestone to be made into a pond' (Dale's characterization). Perhaps the spectre of the dead raccoon will be raised – Dale says it made its appearance on her property around the time the City workers showed up, but that they didn't take it away, despite the City's program to remove

CREATING A BOG GARDEN

With minimal materials and about an hour's work, you can create a bog garden that will help prevent storm-water from overloading Toronto's antiquated sewer system.

The best place for a bog garden is at the base of the downspout that drains water off your roof. If the downspout empties close to your house (say, within a foot), consider buying some kind of extension tubing (vinyl tubing is available from any hard-ware store) and connect it to the downspout in order to move the drain-age water farther away from the house.

Dig a hole that's ap-proximately 12-18 inches deep, with sloping sides, and as large as you want your bog garden to be (it should be a minimum of 2 feet wide), at the downspout base. Line the hole with plastic sheeting, into which you've punctured small holes every 6 inches or so. Cover the lined hole with the excavated soil, and you're ready to plant.

Moisture-loving native species are best. These include sedges such as tussock sedge, wildflowers such as joe-pye weed, swamp milkweed and vervain. You can even get fancy and include carnivorous bog plants such as the pitcher plant, which digests insects the plant traps in its 'pitcher.'

Water well after plant-ing and, if there's no rain, every two days for the next couple of weeks, or during times of drought.

dead animals should a citizen call for the service. Debate will again rage over the aesthetics of someone's gardening choices. Function – the role of a bog in the hydrological cycle, the good that gardens can do – will most likely take a back seat to phantom fears of spontaneously combusting mulch.

One of the stranger lessons of these stories is that the universe loves irony. Before his garden was threatened, Douglas Counter had designed a brochure, published by Etobicoke's Parks and Recreation Services, that urged homeowners to plant boulevard gardens. The week before Deborah Dale's garden was cut down, she had taught a seminar funded by the City and the TRCA's Community Program for Stormwater Management on how to create downspout bog gardens. The ironies pile up: Counter's garden has been featured on numerous television programs, and an image of his garden appeared on an award-winning Canadian postage stamp. Deborah Dale's garden was a finalist in the water-conservation category of the City's 2008 Toronto Green Awards. The convergences don't stop there; Douglas Counter was a director of the North American Native Plant Society for a couple of years and Deborah Dale has been a NANPS director for a decade, including a stint as president. Seems like a dangerous board on which to serve. Indeed, Sandy Bell, the gardener whose own successful court case over a naturalized planting on the front yard of her Beaches home led to the City being forced to provide a 'natural garden' exemption in its Grass and Weeds Bylaw, was a NANPS director in the 1990s. These gardeners are not the only ones being challenged by the City for their naturalized gardens – they just happen to be the ones who have persevered, taken the City to court and, in each case to date, won significant advances for the naturalization movement.

The summer of 2008 brought what seemed like daily storms to the city. Afternoon dramas of dark thunderclouds and crashing downpours scattered people on mad dashes for the protection of porches and store awnings. Streets temporarily became rivers, and raw sewage splattered Sunnyside Beach. Our technologies failed us. Or, rather, we failed to find one of the technologies already at hand – the technology of trowels and plants transforming impermeable surfaces into absorptive sponges that take water where it needs to go. And storms will continue

*Douglas Counter's garden goes postal.*

to brew. What if, instead of fighting these occasional battles caused by the disconnect between City policy, City regulations and innovative individual actions, we rewarded gardeners for wise-water policies? Numerous City departments (Parks, Forestry and Recreation, Toronto Water, Clean & Beautiful City, Toronto Environment Office) already promote naturalized gardens. How about tax breaks for naturalized yards – bucks for bogs and boulevards? Not just safety policies, but energetic promotion of boulevard plantings. Green teams that work with homeowners, school boards and community groups to create demonstration bog gardens, ponds, swales and wet meadows in yards, schools and parks. Toronto has spent millions trying to solve the 'problem' of stormwater, and thousands on court cases over naturalized gardens designed so that water seeps into the ground. How long before this simple idea soaks into the city's consciousness and percolates deep down into a policy of garden protection and, dare we dream, meaningful infiltration- and bog-garden encouragement?

Bert Archer

# Eau de toilette, or how to behave when we're flush

When Lou Di Gironimo, general manager of Toronto Water, sees one of his neighbours watering his sidewalk in the summer, he walks over and gives him a short-form, friendly spiel. He says that, more often than not, his neighbour tells him to piss off: that it's freakin' water, not oil, that there's a huge Great Lake of it in our collective front yard and that even if we went through that one, there are four more Great Lakes where that came from, and that if he wants to water his sidewalk, his aluminum siding, his dog's back and the roof of his freakin' garage, he's gonna go right ahead. Or something to that effect.

I ask this bred-in-the-bone bureaucrat in his windowless City Hall office what his response to this diatribe is, and he grins, because he doesn't have one. His neighbour's right. Even pessimistic aqua-activists have to admit, when pressed, that if the world does fall into a water crisis, we in Toronto have enough to last more than a century, probably two. In fact, if there were a worldwide water shortage at any point in those two centuries, Toronto would probably get rich – and not USA rich, either; Saudi rich, Dubai rich – from selling off its almost endless supply. So why should Di Gironimo care about his neighbour braising his sidewalk?

The answer's simple. Because our per capita consumption of water is eight times what it is in England and Denmark. Because we're using potable water to sprinkle our exotic-grass lawns to the point of drowning them. Because we regularly use as much as twenty litres of water – perfectly clean, perfectly drinkable water, cleaner than the stuff you get in Naya and Dasani bottles – to flush away about 100 grams of our feces. Because there's some truth to what your parents used to say about eating your Brussels sprouts. Though mailing them to the starving children in China was never an option, not appreciating the abundance around you leads to the sorts of habits that create markets for disposable razors and Swiffers. Because the millionaire who lights his cigars with hundred-dollar bills is always a bastard.

But let's say you are a bastard, as many of us are; you should still care, because it's expensive. But not in the way that things are usually expensive. For some ridiculous reason – blame the politics of abundance – this city, like most Canadian cities, has

not been charging us what water actually costs. Despite the fact that we've got loads of it, and it's just right there, water needs to be processed and moved uphill and then moved back down and processed again. For a city of increasing millions, this takes treatment plants, water pipes and sewer systems. Which means two things, one big and one huge. The big one: according to Di Gironimo, the first large-scale sewer systems were built starting in the 1880s, and the first big treatment plants about two decades years later. Toronto, like many other cities, is still using its original ones and, as Di Gironimo says, 'The system life is coming to an end in our lifetimes.'

But it couldn't be happening at a better time. With talk of greenhouse gases, carbon footprints and climate change, we're as environmentally – if not ecologically[1] – sensitive as we've ever been. We're starting to realize, with our increasing urban, upper-class penchant for paying five dollars for a pound of organic tomatoes or an organic, fair-trade chocolate bar, the value of things we previously only knew the prices of. So it's really the perfect time for the City to be slamming us with increased water rates, which is their cute little way of telling us that, over the past century or so, when water was mostly sold for a flat rate based on the number of faucets in a house, they just plum forgot about capital depreciation. Or as Di Gironimo puts it, 'We didn't design the financing model for replacement.'

1 If you want a really good explanation of the vast difference between these two words, please call Councillor Gord Perks at 416-392-7919. He'll be happy to schedule forty-five minutes some Saturday to get into it with you.

Then there's the huge one: in 2004, the last year for which the accumulated numbers are available, the Toronto government used 1.7 billion kilowatt hours of electricity. Of that, the TTC, with its third rails and its streetcars, used 116 million kilowatt hours. Street lighting used 119 million. It took 123 million kilowatt hours to care for all the City's parks and recreational facilities. Toronto Community Housing, the second largest social housing provider in North America, which services 58,500 households, used 421 million. But it took 563 million kilowatt hours to handle our water. That means one out of every three watts used – one third of all electricity used by Canada's biggest city every year – goes to pumping, draining, filtering and treating our water. And it's not all beer-bottling plants and whatnot. It's us. Fifty-one percent of the water used in the city is used residentially. Those kilowatt hours go into treating the water that we flush, let drip from the faucets, waste getting the temperature right and broadcast over our demanding and mostly ungrateful foreign lawns. And yet, the City still charges us a unit rate of $1.70. That's $1.70 for every thousand litres of water we use. We're willing to pay up to three dollars for half a litre of Dasani water, which Coca-Cola freely admits is tap water, and yet the City – and presumably we, too – figure that the most the market can bear for publicly distributed water is 3,500 times less.[2]

Now, and for the foreseeable future, Toronto's water rates will go up by 9 percent a year to cover the costs of repairing and replacing the infrastructure and to pay for some of that electricity,[3] and starting next year in a rollout Toronto Water figures will take about six years, the 70,000 to 72,000 homes that still aren't metered will be. 'Once you meter everything,' Di Gironimo says, 'consumption goes down.'

But Councillor Gord Perks, sitting in his underground Saturday constituency office deep beneath Parkdale Library, points out that the ratio between cost and consumption habits is not one-to-one. 'Witness the purchase of bottled water,' he says. 'The purchase of water is commercially unreasonable.' (See above.)

Perks's point is that habits are at least as acculturated as they are determined by cost. But, aside from the fact that water consumption should drop, Di Gironimo and Perks don't agree on much. They are each a fair representation of the City's sparring aquacultures.

2 For what it's worth, when companies like Coca-Cola sell us tap water in bottles, it's water we've already paid to have treated and pumped. Which either makes Coca-Cola very clever or us really stupid. What's next? Pure Toronto Island air, only fifty cents a puff?

3 And just for fun, raising the price of our water by 9 percent a year, it would take about twenty years of inflation-adjusted rate increases for it to get to Coca-Cola's 2008 rate.

'Recently, we had a meeting about what the key missions of the various departments in the City were,' says Perks. 'The water department said it had two key missions: it was to bring water in, and to flush water out. What their key mission ought to be is to manage the most key nutrient cycle: the water cycle. But they still see it as a series of programs rather than a perspective from which they do their work.'

As far as Di Gironimo's concerned, the most important thing is peak demand. In 2001, the year they decided to set their benchmark, the peak day demand in Toronto and York Region – the day was in August – was 2,211,000,000 litres of water. The system's capacity is 2,200,000,000 litres, but they were able to manage it because there's always water sitting around in reservoirs to make up a little extra. 'The key behaviour we have to modify is peak demand,' Di Gironimo says.

Perks disagrees. He feels we need 'a more balanced relationship with the hydro-geological cycle,' and says that 'we have yet to integrate ecological thinking into our lives. Water is one of the two or three things you really need to understand if you're going to understand life and what its limits are.'

That would usually sound pretty weird coming from a City Councillor. But the fact that councillors like Glenn De Baeremaeker, Paula Fletcher and Pam McConnell talk that way, too, as does the mayor, makes it all a little less Birkenstock and a little more business as usual in the most radically environmentally activist City Council we've ever had. But, as we've seen, it also makes for some run-ins with the bureaucrats who figure these guys are nuts. And since the population at large would also, for the most part, take poorly to a city that demanded they rethink their twenty-minute showers in order to understand life and what its limits are, the synthesis of the two hydropolitical visions, hammered out in what one imagines are probably often fractious meetings, seem to be working pretty well.

Take toilet subsidies, for instance. Though installing a low-flow toilet does appreciably improve one's relationship with the hydrogeological cycle, people mostly do it for the $60 they get back. At the end of 2007, the City had handed out 216,749 low-flow toilet subsidies, and of the 175 megalitres of water savings they've realized from all sources, 101 are from the reduction in flushes. By the end of 2008, the City expects to save 50 million litres of water per day as a result of these and the high-efficiency washing machines, which they've also been pushing (28,021 subsidies for those had been handed out by the end of 2007). You can also call the City, or stop by the Access Toronto desks at City Hall or the Etobicoke, North York and Scarborough civic centres, and pick up what they're calling an indoor water-efficiency retrofit kit (I-WERK might be a nifty acronym). It includes one high-efficiency showerhead that saves 40 percent of the water an average head uses, one kitchen tap aerator, two bathroom aerators, a package of leak-detection tablets to put in your toilet tank to see if it's leaking into your toilet bowl, and a roll of Teflon tape to prevent leaks in your shower arm. But the City, being in the municipal politics game and seemingly lacking in sales experience, has managed to sell, as of this writing, exactly 500 of them since the program began in 2004. So let me offer them some assistance.

> *Now, you might think you'd have to pay more than $50 for this incredible collection of money- and planet-saving devices in a store, but because you pay your taxes, and because water costs the City so much damn money, it's available to you in this unlimited-time offer for less than you paid for this book. That's right. You can be part of a fraternity more exclusive than the York Club or the Andrew Dice Clay fan club for the low, low price of $10. Can you afford not to? Of course you can't. The secret number to call is printed below in a footnote.*[4] *Call now!*

Still, Di Gironimo worries about the peak. He worries because the City's infrastructure has to be built up and able to handle peak demand, even if it only comes one day a year. And peak water, just like peak electricity, happens in the summers. Of course, we still poo in the summers, and we may even wash our

4 416-338-0338. That's 416-338-0338. Operators are standing by.

clothes slightly more often, with kids' grass stains and adults' sweatiness at their peak, too. But water's equivalent of air conditioning, which David Suzuki and his penguins remind us is the summer's electrical arch-villain, are lawns, with a little bit of extra car-washing mixed in.[5] And since there's no such thing as a low-flow hose yet (I don't think), the City's put together a team of consultants who can drop by your house and advise you on the proper use of water in landscaping. One of the things they'll recommend is how great rock gardens are, needing just a fraction of the water a regular lawn needs. One of the things you won't hear about, though, are prairie lawns. Apparently, people think they look messy, and have some problems convincing themselves that native plants aren't just icky weeds. Ah well, maybe next year.

Rock gardens and giving our poop a little less of the water-park experience are both parts of the Water Efficiency Plan put together in 2002 and approved in 2003. According to Di Gironimo, 'One of the big drivers was trying to ensure that if we used less, we could retain the infrastructure that we had and service more people. It was about stretching the infrastructure.' But what they found, once they started to implement this plan – one of the more ambitious North American ones outside of places like Georgia and California, where water is an annual problem – was that people were way more willing to comply than they'd thought. It turns out that even we didn't want to use as much water as we were using. Goals that seemed ambitious four years ago have been exceeded to such an extent that Toronto Water is rejigging the whole plan to take this unanticipated enthusiasm into account. Ave Maria.

But according to Di Gironimo, this rejigging is one of degree, not kind. There are, as he said, still no plans to talk prairie lawns. As far as the water guys are concerned, the good news is

5 Cars must be good for something; watch for *auTOpia* as soon as the kids at Coach House have discovered what.

that the infrastructure will now last longer that they'd thought, and that peak has been reduced by more than they had expected, which means we can take more people using these amounts of water before those 9 percent water-rate increases have to turn into 18 percent increases to fund the construction of entirely new treatment plants and bigger sewers. At current rates of consumption and degree of reduction, the best-before date for our water system is a very respectable 2035.

One distinctly Perksian project that's in the works in south Etobicoke is the Silva Cell, produced by urban landscape products supplier DeepRoot. A set of plastic cube frames planted underneath water-permeable sidewalks or other paving, it's meant to maintain the soil inside the frame uncompacted and relatively loose, allowing plant roots to get in there and take first pull at stormwater before it flows off into the sewer system. The plan is to test how much water ends up being diverted up from the sewers and into roots and branches, and at what cost.

Perks is also talking about incorporating green space into parking lots to give water a natural, rather than mediated and engineered, place to go when it falls from the sky. He credits people like Shelley Petrie, formerly of the Toronto Environmental Alliance, and Kevin Mercer from RiverSides with lobbying efforts that have gotten the City to start to change its way of thinking. 'We're trying to get stuff added to HTO Park[6] to run a living machine,' Perks says. 'You use some features of a wetland to treat water – you can even theoretically use it to treat sewage.'

As far as Perks is concerned, there's a difference between efficiency and conservation, though they're not mutually exclusive. One can be done with retrofit kits; the other requires, through great feats of engineering and personal rethinking of things as basic as how often we wash our hair and how much effort we should rightly expend to do something useful with all that downspout water we're collecting, a reintegration of this great big metal and concrete machine we live inside into the water cycle that's been battering away at it, trying to jog our memories with every rainfall that floors our sewers and every snowstorm that creates those big lakes of melting water that stall our cars, that as soon as we stop working against it, it can begin to work with us.

6 The park, at 339 Queen's Quay West, funded with money put aside by Prime Minister Pierre Trudeau and opened in June 2007. The pace at which this project moved has been called glacial, but that's not fair – an average glacier would have moved more than five kilometres in the time it took to make this 2.3 hectare park.

# Contributors

RICHARD ANDERSON teaches geography part-time at York University. He is a member of the Toronto Green Community's Lost Rivers Committee and researches the historical geography of the Toronto environment. He is preparing a book on the historical geography of Toronto's atmosphere.

BERT ARCHER is a writer about many things for *Toronto Life*, the *Globe* and others. He lives in the Annex and works in cafés and pubs from the Junction (Crema) to Bloor and Bathurst (Aroma) to Greenwood (the Hargrave). He hears Stephen Marche hangs out at the Common, so he will have added that to the list by the time you read this. He likes water well enough, though figures that no writing can live long or please that is written by water-drinkers.

CHRIS BILTON is a freelance journalist living and working in Toronto. He is a regular contributor to *Eye Weekly*, writing for both the music and city sections, along with a number of other local and international publications such as *Spacing*, *Ukula* and *Slash Magazine* in New York, for which he is a contributing editor.

JENNIFER BONNELL is currently completing a PhD in the History of Education Program at the University of Toronto's Ontario Institute for Studies in Education. Her dissertation research explores the social and environmental history of the Don River in Toronto, and the ways social, political and economic motivations, as well as ecological forces themselves, have shaped the river's course and condition – and people's responses to it – over time.

JAMES BROWN and KIM STOREY are partners in the architectural and urban design firm of Brown and Storey Architects. Over the past twenty-five years, the pursuits of B+SA have ranged from a broad landscape urbanism to scenarios of public space, built form and infrastructure. Office projects include Yonge-Dundas Square, Massey Harris Park, Garrison Creek and College Street, and the firm has led and been on international teams for the Lower Don Lands and the Downsview Park competitions. The Garrison Creek work has been exhibited in the Venice Biennale and published internationally, as have many other office projects.

MICHAEL COOK is a graduate student at York University, where his research focuses on the human geography of resource exploitation in remote regions. Cook's work exploring and photographing the underground infrastructure of sewerage and hydroelectric power generation in Ontario is documented on his website, www.vanishingpoint.ca.

NICK EYLES has been professor of geology at the University of Toronto since 1981 and, in addition to ongoing research, has written several best-selling books on the geological evolution of Ontario and Canada. He was scientific advisor to CBC's recent *Geologic Journey*, a five-part series on Canada's long geologic history that aired in late 2007.

MARK FRAM has trained as an architect and as a geographer, putting those disciplines to work over the years (decades!) in urban design, environmental planning, historic-building preservation, scholarly research and professional education. At Coach House, he is a member of the select group of *clavigeri*, and from time to time applies himself therein variously as essayist, editor, architectural consultant, typographical aficionado and designer of printed things.

ED FREEMAN, trained in geography and geology, spent twenty-seven years as a mineral scientist and public-information specialist with the Ontario Geological Survey, eight years teaching geography in secondary school and has led many earth-science–focused field trips over the last forty years. He is a member of four Toronto-based historical societies and is active in leading walks in Toronto for Heritage Toronto, the Lost Rivers group and the Toronto Field Naturalists.

LIZ FORSBERG was born on the banks of the Speed River in Guelph, Ontario. She is a community artist, musician and writer with a Master's degree specializing in the intersection of art, ecology and public space. She has developed, coordinated and facilitated numerous arts-based educational projects in collaboration with the Art Gallery of York University, the Toronto Public Space Committee, Lost Rivers, Local Enhancement and Appreciation of Forests and the Department of Geography at York University.

MICHAEL HARRISON has long been interested in the history of south Etobicoke. He was a founding member of Citizens Concerned About the Future of the Etobicoke Waterfront and was its president for ten years. In 1997, he researched and wrote *Toward the Ecological Restoration of South Etobicoke*, which examines the environmental history of the area, and proposed a number of restoration projects, some of which have already been implemented. He was a founding member of the New Toronto Historical Society and has been actively researching the fascinating people who lived in the waterfront estates of the Town of Mimico in the late nineteenth and early twentieth centuries.

As an associate at Sweeny Sterling Finlayson & Co Architects Inc., CHRIS HARDWICKE researches, designs, teaches, gives advice, makes policies and writes about places and cities. His visionary urban projects have been exhibited and published widely, including in the books *uTOpia* and *GreenTOpia* from Coach House Books.

MAGGIE HELWIG is the author of six books of poetry, two books of essays, a collection of short fiction and three novels, most recently *Girls Fall Down*. She lives in Toronto and walks around a lot.

LORRAINE JOHNSON writes about gardening and its connection to environmental and social issues. Her file on gardeners who have been harassed for their naturalized plantings is four inches thick, and grows bigger each summer.

JOANNA KIDD is a writer, environmental consultant and long-time Toronto Island resident. She has written a number of books relating to the island and Toronto Bay, including *A Living Place: Opportunities for Habitat Restoration in Toronto Bay* and *Nature on the Toronto Islands: An Explorer's Guide*. Joanna has worked as an environmental consultant since 1988 and has delivered many water-related projects for municipal, provincial, federal and international clients. She has sailed, windsurfed, paddled and rowed on the waters of Toronto Bay since 1966 and was a founding board member of the Toronto Bay Initiative, and later its chair.

JOHN LORINC is a Toronto journalist who covers municipal affairs for *Spacing* and the *Globe and Mail*. He is the author of *Cities: A Groundworks Guide* (Groundwood Books, 2008).

**ROBERT MACDONALD** is a partner and senior archaeologist with Archaeological Services Inc. and an adjunct assistant professor in the Department of Anthropology at the University of Waterloo, as well as a research associate of the Trent University Archaeological Research Centre. His special areas of expertise include Iroquoian archaeology, ecological archaeology, archaeological site potential modelling and geographic information system applications in archaeology.

**STEVEN MANNELL, NSAA, MRAIC,** is a professor at the School of Architecture at Dalhousie University and principal of Steven Mannell, Architect, in Halifax. He has published several academic articles on the Toronto Water Works Extension, and has been active in heritage advocacy for the R. C. Harris site. His connection to the St. Clair Reservoir site is more intimate: years before beginning this research, his and Laureen's wedding portrait was taken in front of the pipe-tunnel portal. His essay is based on research funded by the Social Sciences and Humanities Research Council of Canada.

**MICHAEL MCMAHON** curated the critically acclaimed *Pipe Dreams: A History of Water and Sewer Infrastructure in Metropolitan Toronto* exhibit at the Metro Toronto Archives in the mid-nineties. The Metro Works Department provided support for a related virtual exhibit, available at www.toronto.ca/archives/pipedreams/splash.htm. Michael is completing a doctoral degree in environmental studies at York University.

**SHAWN MICALLEF** is associate editor of *Spacing* magazine and co-founder of [murmur], the location-based mobile-phone documentary project. He writes about cities, culture, buildings, art and whatever is interesting in various books, magazines and newspapers.

**GARY MIEDEMA** is a historian with Heritage Toronto. A regular contributor to *Spacing* magazine, he also teaches Canadian history in Ryerson University's School of Continuing Education.

Entrepreneur-turned-environmentalist **HELEN MILLS** has spent the past twelve years engaged in environmental activities at the community level. In 1995, with Peter Hare, Ed Freeman, Julie Nettleton and James Brown, she started the Lost Rivers Project of the Toronto Green Community. A deep interest in plants and natural history led her to community gardening, to starting the Green Garden Visit and now to Green Gardeners Community Collaborative Inc., which is all about changing the urban grid, one garden at a time.

A proud Canadian born in East Africa, MAHESH PATEL lived and was educated in England, but found his passion for environmental public health and an ideal 'meeting place' in Toronto. While working for one of the best cities in the world, he has been inspired to, in the words of a truly great Canadian, 'truly know his passion and to live it.'

FRANK REMIZ has helped advance sustainable transportation and urban greening initiatives in the boardroom and on the ground for twenty years. He works as a human-resources consultant.

RIVERSIDES' essay is a summary of a larger report, co-authored and researched by staff and volunteers, including EMILY ALFRED, JUSTYNA BRAITHWAITE, JP WARREN, KEVIN MERCER, JENNIFER DILLON and MARIKO UDA. Research for the report was funded by the Toronto Atmospheric Fund.

DAVID A. ROBERTSON is a senior archaeologist at Archaeological Services Inc., a cultural resource management firm based in Toronto. Much of his work is focused on the historic urban core and waterfront of the city.

JANE SCHMIDT is co-chair of the High Park Volunteer Stewardship Program. She works for a museum and is a freelance writer. She was the main author of *Rare Plants of the Endangered High Park Black Oak Savannah* (2008).

Former Toronto resident MURRAY SEYMOUR has been exploring Toronto's ravine lands for over thirty years. He is author of *Toronto's Ravines: Walking the Hidden Country* (2000).

EDUARDO SOUSA works as a consultant in community planning and sustainability. He was born in Portugal, by the Atlantic Ocean, but lived in Toronto, near Lake Ontario, for most of his life. Eduardo has a Master's degree in Environmental Studies from York University in watershed education and planning, and he has worked for City of Toronto Parks and the Council of Canadians. Despite his love of fresh water, the ancient allure of the ocean has called him to head west. He now makes his home in Vancouver, where he intends to reinhabit those watersheds.

ANDREW M. STEWART is a consulting geo-archaeologist interested in the evolution of landscape and is chair of the History and Archaeology Committee of the Friends of Fort York and Garrison Common.

**RONALD F. WILLIAMSON** is managing partner of Archaeological Services Inc., the largest archaeological consulting firm in Canada. He is also an associate of the Graduate Faculty at the University of Toronto and is published widely in both academic and commercial presses. He is the editor of *Toronto: A Short Illustrated History of Its First 12,000 Years*, in which he also authors a chapter on the Aboriginal history of Toronto.

**GEORGIA YDREOS** is a Pisces, at home swimming in rivers, lakes, oceans and, less thoroughly, swimming pools. She has a fervour for cultivating and integrating upstream and downstream ecologic consciousness, and applies it to multidisciplinary projects that span fine and graphic arts, interior and architectural design, community art and advocacy. Her experience co-organizing Human River with Erin Wood and many dedicated Toronto Public Space Committee volunteers (including co-writer Liz Forsberg) has provided a profound underpinning for her University of Waterloo graduate architectural thesis entitled 'An Architecture Drawn on Water.'

**CHRISTINA PALASSIO** is the managing editor of Coach House Books. She co-edited, with Johnny Dovercourt and Alana Wilcox, *The State of the Arts: Living with Culture in Toronto* and *GreenTOpia: Towards a Sustainable Toronto*. She has also written for the *Globe and Mail*, the *Montreal Gazette* and *Matrix Magazine*.

**WAYNE REEVES** is a project officer with Toronto Parks, Forestry & Recreation who prefers his water with malt, hops and yeast. His previous publications include two U of T research monographs on the Toronto waterfront, several heritage studies (including one on the R. C. Harris Water Treatment Plant) and essays in *Special Places: The Changing Ecosystems of the Toronto Region* and *GreenTOpia*. He's currently leading the development of the Toronto Beaches Plan.

# Illustrations: Sources & credits

**Pages 2-3** *Lifeguard competition at Woodbine Beach, 2007. Photo by George Leet. Courtesy of the City of Toronto.*

**8-9** *Laying a water intake for the deep-lake water-cooling project, 2004. Courtesy of Enwave.*

**10** *Sunnyside Free Bathing Station on Lake Ontario, c. 1907. Photo by William James. City of Toronto Archives, Fonds 1244, Item 222.*

**10** *Dunnett's swimming hole on the West Don River, west of Bayview Avenue, c. 1900. Photo by William W. Judd. Toronto Reference Library, Baldwin Room, 978-13-11.*

**11** *Kidstown waterplay facility at L'Amoreaux Park, east of Birchmount Road, 2008. Photos by Gera Dillon. Courtesy of the City of Toronto.*

**13** *Photo by Wayne Reeves, 2008.*

**15** *[top] Photo by Wayne Reeves, 2008.*

**15** *[bottom] City of Toronto, Museum Services, Colborne Lodge Collection, 1978.41.30.*

**17, 19** *Photos by Wayne Reeves, 2008.*

**21** *[top] Photo by Wayne Reeves, 2008.*

**21** *[bottom] Courtesy of the City of Toronto.*

**23** *Drawing of the now-extinct giant beaver by Charles Douglas, reproduced with permission of the Canadian Museum of Nature, Ottawa, Canada.*

**24** *Rouge River at the Finch Meander, 2008. Photo by Wayne Reeves.*

**24** *Native Americans making a dugout canoe, by Theodore Bry after a sixteenth-century John White watercolour. One such canoe was found in Bond Lake in York Region. Courtesy of Archaeological Services Inc.*

**24** *Joseph Bouchette, Plan of York Harbour, 1815. Toronto Reference Library, Baldwin Room, Book Collection.*

**25** *Scarborough Bluffs, 2007. Photo by John Davidson. Courtesy of the City of Toronto.*

**27** *Drawings by Ed Freeman.*

**30** *Diagram by Peter Russell. Courtesy of Paul Karrow, University of Waterloo.*

**31** *Photo by Nick Eyles.*

**33** *Photo by Ed Freeman.*

**35** *Map taken from the* Canadian Journal of Economics and Political Science, *Vol. 2, No. 4 (November, 1936), pp. 493-511.*

**36** *Maps and diagram by Nick Eyles.*

**37** *Map by Nick Eyles.*

**39** *Photo by Nick Eyles.*

**40 [6]** *Photo by Nick Eyles.*

**40 [7]** *Diagram by Nick Eyles.*

**45** *Painting by Shelly Huson. Courtesy of Archaeological Services Inc.*

**46, 47, 48** *Courtesy of Archaeological Services Inc.*

**49** *Service historique de la Marine, Bibliotheque, Paris.*

**51** *Photo by John Howarth. Courtesy of Archaeological Services Inc.*

**52** *Drawing from the* Canadian Illustrated News *(June 10, 1871).*

**53** *Maps adapted by Chris Hardwicke from the City of Toronto Planning & Development Department's* City Patterns: An Analysis of Toronto's Physical Structure and Form, Cityplan '91 Report 29 *(Toronto, 1992).*

**54** *Map by Chris Hardwicke, based on Toronto and Region Conservation Authority data.*

**56** *Map from the* Report to the City Council of Toronto on the Proposed Water Supply by Gravitation from the Oak Ridge Lakes and the Rivers Don and Rouge, *by William J. McAlpine and Kivas Tully (Toronto: E.F. Clarke, 1887).*

**58, 59** *Maps by Chris Hardwicke.*

**60** *Map by C. W. Jeffreys. From* Toronto During the French Régime: A History of the Toronto Region from Brûlé to Simcoe, 1615-1793, *by Percy J. Robinson (Toronto: Ryerson Press, 1933; revised edition, 1966).*

**61** *City of Toronto Archives, Series 725, File 12.*

**62** *Map from the* Indian Treaties, Copy Book of Deeds & Provisional Agreements by the Chiefs of the Chippewa and Mississauaga Tribes (1811). *Toronto Reference Library, Baldwin Room, Manuscript Collection.*

**63** *Plan of Todmorden Mills, 1855. Courtesy of the City of Toronto.*

**64** *Construction of the Ship Channel in Ashbridge's Bay, September 25, 1919. Photo by Arthur Beales. Toronto Port Authority Archives, PC 1 / 1 / 4304.*

**64** *Woodville Avenue dump, June 11, 1914. Photo by Arthur Goss. City of Toronto Archives, Fonds 1231, Item 1459.*

**64** *Drawings from* Water Works and Sewerage Systems of Canada, *by Leo G. Denis (Ottawa: Canada Commission of Conservation, 1916).*

**65** *Circa 1940 drinking fountain and urinal at R. C. Harris Water Treatment Plan, 2008. Photo by Wayne Reeves.*

**65** *Black Creek Channel near Jane & Wilson, 2008. Photo by Wayne Reeves.*

**65** [below] *Alexander Street sewer construction, 1949-50. Photo by Canada Pictures. City of Toronto Archives, Series 381, File 76, Item 7427-13.*

**67** *City of Toronto Archives, Fonds 1231, Item 302.*

**69** *Courtesy of the North York Historical Society.*

**70** *Drawing from* Illustrated Toronto: Past and Present, *by James Timperlake (Toronto, 1877).*

**72** *From hill above Belt Line on the Humber, showing the Old Mill and Bloor Street Bridge* [*between 1895 and 1900*]. *Archives of Ontario, C157-0-0-0-49.*

**75** *Map by Richard Anderson.*

**76** *Photo by Wayne Reeves.*

**79** [**3**] *City of Toronto Archives, Series 4, Sub Series 2, item 457.*

**79** [**4**] *City of Toronto Archives, Series 4, Sub Series 2, item 461.*

**81** *Map by Richard Anderson.*

**83** [**1, 2**] *Courtesy of the Toronto and Region Conservation Authority.*

**85** *Courtesy of the Toronto and Region Conservation Authority.*

**86-87** *Photos by Wayne Reeves, montage by Mark Fram.*

**88** *Photo by Steven Evans.*

**91** *Drawing by Michael Van Valkenburgh Associates, Inc. Courtesy of Waterfront Toronto.*

**93** *Toronto Reference Library, Baldwin Room, Canadian Historical Picture Collection.*

**95** *City of Toronto Archives, Series 376, File 3, Item 9d.*

**97** *City of Toronto Archives, Fonds 200, Series 376, File 1, Item 33a.*

**98** *City of Toronto Archives, Series 376, File 1, Item 29b.*

**99** *Photo by William James. City of Toronto Archives, Fonds 1244, Item 6034.*

**99** *Diagram by Mahesh Patel.*

**101** *Photo by Arthur Goss. City of Toronto Archives, Fonds 1231, Item 780.*

**104** *Diagram from the* Contract Record and Engineering Review, *(June 19, 1935).*

**106** *Courtesy of* CH2M HILL *Canada Ltd.*

**107** [**3, 4**] *Digital models and renderings by Steven Mannell and Chad Jamieson, 2003.*

**108** [**5, 6**] *Photos by Arthur Goss. Courtesy of the City of Toronto.*

**109** [**7**] *Photo by Arthur Goss. Courtesy of the City of Toronto.*

**111** [**8**] *Drawing from the* Engineering Journal *(February 1935).*

**111** [**9**] *Digital model and rendering by Steven Mannell, Benjamin Checkwitch and William Rawlings.*

**112-113** *Base photos by Arthur Goss, 1934. Digital model and rendering of tank and photomontage by Steven Mannell, Benjamin Checkwitch and William Rawlings, 2000.*

**113** [**11**] *Copyright unknown. City of Toronto Archives, Series 4, Subseries 1, Item 65.*

**115** *City of Toronto Archives, Series 372, Subseries 36, Item 57.*

**116** *City of Toronto Archives, Series 4, Subseries 1, Item 264.*

**118** *Copyright unknown. City of Toronto Archives, Series 160, File 130.*

**119** [**4**] *City of Toronto Archives, Series 4313, Subseries 9, File 1, Item 6.*

**119** [**5**] *City of Toronto Archives, Series 4, Subseries 1, Item 130.*

**120** *Photo by Nick Shinn.*

**123** *Miles & Co.,* Illustrated Historical Atlas of the County of York *(Toronto, 1878).*

**125** *Photo by William James. City of Toronto Archives, Fonds 1244, Item 2481.*

**126** *William Sims, watercolour over pencil. Royal Ontario Museum, catalogue no. 952.152, Yorkville Waterworks, ROM2008_10284_1.*

**128** *City of Toronto Archives, Series 376, File 5, Item 76.*

**129** *Photo by Arthur Goss. City of Toronto Archives, Series 372, Sub-series 1, Item 670.*

**130** Goad's Atlas of Toronto, *1923, Plate 38. Toronto Reference Library, Baldwin Room, Book Collection.*

**133** *Scarborough Bluffs, c. 1909. Photo by William James. City of Toronto Archives, Fonds 1244, Item 1537.*

**134** *Sunnyside Beach, April 6, 1928. Photo by John Boyd. Library and Archives Canada, PA 088011.*

**134** *Photo by Wayne Reeves, 2008.*

**134** [*lower*] *'Toronto Sewerage System - One of the Mains.' Photo from* Water Works and Sewerage Systems of Canada, *by Leo G. Denis (Ottawa: Canada Commission of Conservation, 1916).*

**135** *High Park, c. 1910. Courtesy of the City of Toronto.*

**135** *Tommy Thompson Park, 2007. Photo by George Leet. Courtesy of the City of Toronto.*

**135** [*lower*] *North Toronto Storm Trunk Sewer. Photo by Michael Cook.*

**136-137** *Toronto Water Works,* Annual Report of the Board of Water Commissioners for *1875. Toronto Reference Library.*

**138** *Photo by Galbraith Studios. Courtesy of the City of Toronto.*

**140** *City of Toronto Archives, Series 163, File 108, Item 3.*

**141** *City of Toronto Archives, Series 163, File 108, Item 1.*

**142** *Photo by Edwin Luk.*

**144** *City of Toronto Archives, Series 163, File 69, Item 1.*

**145** [7] *City of Toronto Archives, Series 163, File 108, Item 2.*

**145** [8] *Photo by Wayne Reeves.*

**147, 149, 150, 151, 152** *Courtesy of Archaeological Services Inc.*

**155** *Map courtesy of the Toronto and Region Conservation Authority.*

**156** *Detail from a map by H. J. W. Gehle. Library and Archives Canada, NMC 26685.*

**158** *Postcard by Heroux Postcard Manufacturers, Trois-Rivières, QC. Collection of Michael Harrison.*

**159** *Canadian Underwriters Association, Underwriters' Survey Bureau Ltd.,* Insurance Plan of the City of Toronto. *Toronto Reference Library, Baldwin Room, Book Collection.*

**160-161** *Photos by Michael Harrison.*

**163** *Photo by John Boyd. Archives of Ontario, C 7-1-0-0-71.*

**165-169** *All photos by Murray Seymour except the frog on p. 168, which is courtesy of the City of Toronto.*

**171** *Photo by HiMY SYeD (www.photopia.tyo.ca).*

**172** *Photo by Kateryna Topol.*

**173** *Photo courtesy of Clay and Paper Theatre.*

**174** *Photo by Paola Giavedoni (www.paolagiavedoni.com).*

**175** *Photo by Cylla Von Tiedemann.*

**176** *Photo by Cheryl Rondeau.*

**180-182** *Photos by David Barker Maltby.*

**185** *Wilson Heights Storm Trunk Sewer. Photo by Michael Cook.*

**193-200** *Photos by Michael Cook.*

**201** *Drawing by Michael Van Valkenburgh Associates, Inc. Courtesy of Waterfront Toronto.*

**202** *Banjo on the low-flow Toto, 2008. Photo by Wayne Reeves.*

**202** *Poster courtesy of RiverSides.*

**202** *Toronto tap-water bottle, 2008. Photo by Mark Fram.*

**202** [*below*] *Toronto Water ad, 2005. Courtesy of the City of Toronto.*

**203** *Laying a water intake pipe for the deep-lake water-cooling project, 2004. Courtesy of Enwave.*

**203** *Wetland stewardship workshop at Don Valley Brick Works, 2007. Photo by Cheryl Post. Courtesy of the City of Toronto.*

**205** *Courtesy of Brown and Storey Architects.*

**206** [2] *City of Toronto Archives, Fonds 1231, Item 1543.*

**206** [3] *City of Toronto Archives, Fonds 1231, Item 1615.*

**207** *City of Toronto Archives, Fonds 1231, Item 1917.*

**208** *Trinity College Archives P1193/0001.*

**209** *Brown and Storey Architects, Venice Biennale, 1996.*

**210-211** *Brown and Storey Architects.*

**213** *Photo by Wayne Reeves.*

**215** *Photo by Ed Freeman.*

**219** *Photo by Gera Dillon. Courtesy of the City of Toronto.*

**223** *Photo by Wayne Reeves.*

**227, 228** *Courtesy of the City of Toronto.*

**229** *Photographer unknown. Photos courtesy of jane-finch.com (jane-finch.com/pictures/flood2005.htm).*

**231** *Photo by Wayne Reeves.*

**232** *[top] Photo by Mark Fram.*

**232** *Photo by George Leet. Courtesy of the City of Toronto.*

**235** *University of Toronto Archives, Archibald Byron MaCallum, B1966-0005/003(09).*

**236-242** *Maps by Evan Munday.*

**239** *Archives of Ontario, F507ST353.*

**241** *Courtesy of a private collection.*

**242, 243** *Toronto Reference Library, Baldwin Room, Manuscript Collection.*

**244** *Photo by John Boyd, June 17, 1916. Library and Archives Canada, PA 0069864.*

**245** *Photo by John Boyd, June 17, 1916. Library and Archives Canada, A 069865.*

**251** *Courtesy of the City of Toronto.*

**253** *Photo by George Leet. Courtesy of City of Toronto.*

**255** *Map by Toronto Harbour Commissioners, 1912. Courtesy of University of Toronto Libraries.*

**257, 259, 261** *Photographic montages by Mark Fram, 2008. Adapted from 2005 aerial imagery. Base data courtesy of Toronto Public Library.*

**258** *Extracts, National Topographic System 30M/11 [sheet 34] Toronto, 1:63,360: 1909, 1918, and 1931. Department of National Defence. Courtesy of National Map Collection.*

**259** *City of Toronto Archives, Series 376, File 1, Item 43b.*

**262** *Photo by Mark Fram, 2008.*

**263** *Algiers – depot and station grounds of Algerian Railway, 1894. Photo by William Henry Jackson. Library of Congress Prints and Photographs Division, Digital ID 3b25939.*

**264** *Church Street south of Esplanade (formerly Toronto Harbour), looking north to Front Street, 2008. Photo by Mark Fram.*

**265** *Fish Market, Toronto, before 1842 (from the south and the east). Drawn by W. H. Bartlett, engraved by E. J. Roberts.* Canadian Scenery Illustrated *(London, 1842).*

**267** *City of Toronto Archives, Series 80, File 8.*

**269** *City of Toronto Archives, Series 101, File 20.*

**271** *Photo from* The Globe and Mail, *May 30, 1955. Archives of Ontario, RG1-112-1-0-8.*

**272** *Photo by Tom Davey.*

**275** *Photo by Keri McMahon. Courtesy of the City of Toronto.*

**276** *Courtesy of the City of Toronto.*

**281** *Drawing by Michael Van Valkenburgh Associates, Inc. Courtesy of Waterfront Toronto.*

**285** [1] *Photo by Frank Remiz.*

**285** [2] *City of Toronto Archives, Fonds 1231, Item 292.*

**286** *City of Toronto Archives, Series 724, File 22, Item MT 1052.*

**287** [4] *Courtesy of John Pries,* CH2M HILL.

**287** [5] *City of Toronto Archives, Fonds 1244, Item 455.*

**288** [6, 7] *Courtesy of the City of Toronto.*

**293** *Courtesy of* SAS *Canada.*

**294** *Diagram by RiverSides.*

**295** *Drawing by Evan Munday.*

**296** *Photo by Wayne Reeves.*

**304, 305** *Photos by Andrew Leyerle.*

**306, 307** *Photos by Deborah Dale.*

**309** *© Canada Post Corporation (2006). Reproduced with permission.*

**311** *Toronto Water ad, 2005. Courtesy of the City of Toronto.*

**312** *Toronto Water ad, 2006. Courtesy of the City of Toronto.*

**313** *Toronto Water ad, 2005. Courtesy of the City of Toronto.*

**315** *Toronto Water ad, 2001. Courtesy of the City of Toronto.*

**328** *Photos by Mark Fram and Rick/Simon.*

**$H_TO$: Toronto's Water from Lake Iroquois to Lost Rivers to Low-flow Toilets**
**Edited by Wayne Reeves and Christina Palassio**

THE BOOK

$H_TO$ was designed, composed and installed on these pages by Mark Fram, with the peerless help of Stan Bevington and Rick/Simon at the Coach House. Toronto tap water was an essential adjunct throughout production, supplemented on occasion by a bit of rain. Neither water tables nor aquatic life were harmed.

The text types comprise selected varieties of the Freight family, designed by Joshua Darden. The main titles are set in a mildly fuzzy instance of BeoSans, designed by Just van Rossum. Other letterforms make cameo appearances.

$H_TO$ is printed on Rolland ST30, containing FSC-certified 30 percent post-consumer and 70 percent virgin fibre. It is Certified EcoLogo and FSC Mixed Sources and manufactured using biogas energy.

THE COVER

*Untitled*, by Ilyana Martinez (detail).
Watercolour and ink on paper, 28 x 38 cm.
www.ilyanamartinez.com

Cover design by Mark Fram.

**Coach House Books**
**401 Huron Street on bpNichol Lane**
**Toronto, Ontario M5S 2G5**

**416 979 2217**
**800 367 6360**

**mail@chbooks.com**
**www.chbooks.com**